The UN Millennium Project is an independent advisory body commissioned by the UN Secretary-General to propose the best strategies for meeting the Millennium Development Goals (MDGs). The MDGs are the world's targets for dramatically reducing extreme poverty in its many dimensions by 2015—income poverty, hunger, disease, exclusion, lack of infrastructure and shelter—while promoting gender equality, education, health, and environmental sustainability.

The UN Millennium Project is directed by Professor Jeffrey D. Sachs, Special Advisor to the Secretary-General on the Millennium Development Goals. The bulk of its analytical work has been carried out by 10 thematic task forces comprising more than 250 experts from around the world, including scientists, development practitioners, parliamentarians, policymakers, and representatives from civil society, UN agencies, the World Bank, the International Monetary Fund, and the private sector. The UN Millennium Project reports directly to UN Secretary-General Kofi Annan and United Nations Development Programme Administrator Mark Malloch Brown, in his capacity as Chair of the UN Development Group.

Prescription for healthy development: increasing access to medicines

Lead authors
Beryl Leach
Joan E. Paluzzi
Paula Munderi, Coordinator

UN Millennium Project
Task Force on HIV/AIDS, Malaria, TB, and
Access to Essential Medicines
Working Group on Access to Essential Medicines
2005

earthscan
from Routledge

MillenniumProject

First published by Earthscan in the UK and USA in 2005

For a full list of publications please contact:
Earthscan
2 Park Square, Milton Park, Abingdon, Oxon OX14 4RN
52 Vanderbilt Avenue, New York, NY 10017

First issued in hardback 2018

Earthscan is an imprint of the Taylor & Francis Group, an informa business

A catalogue record for this book is available from the British Library

Library of Congress Cataloging-in-Publication Data

A catalog record has been requested

This publication should be cited as: UN Millennium Project. 2005. *Prescription for Healthy Development: Increasing Access to Medicines.* Report of the Task Force on HIV/AIDS, Malaria, TB, and Access to Essential Medicines, Working Group on Access to Essential Medicines.

Photos: Front cover, Curt Carnemark/World Bank; back cover, top to bottom, Christopher Dowswell/UNDP, Pedro Cote/UNDP, Giacomo Pirozzi/Panos Pictures, Liba Taylor/Panos Pictures, Jørgen Schytte/UNDP, UN Photo Library, Giacomo Pirozzi/UNICEF, Curt Carnemark/World Bank, Pedro Cote/UNDP, Franck Charton/UNICEF, Paul Chesley/Getty Images, Ray Witlin/World Bank, Pete Turner/Getty Images.

This book was edited, designed, and produced by Communications Development Inc., Washington, D.C., and its UK design partner, Grundy & Northedge.

The UN Millennium Project was commissioned by the UN Secretary-General and sponsored by the United Nations Development Programme on behalf of the UN Development Group. The report is an independent publication that reflects the views of the members of the Working Group on Access to Essential Medicines of the Task Force on HIV/AIDS, Malaria, TB, and Access to Essential Medicines, who contributed in their personal capacity. This publication does not necessarily reflect the views of the United Nations, the United Nations Development Programme, or their Member States.

ISBN 13: 978-1-138-47183-2 (hbk)
ISBN 13: 978-1-84407-227-9 (pbk)

Foreword

The world has an unprecedented opportunity to improve the lives of billions of people by adopting practical approaches to meeting the Millennium Development Goals. At the request of UN Secretary-General Kofi Annan, the UN Millennium Project has identified practical strategies to eradicate poverty by scaling up investments in infrastructure and human capital while promoting gender equality and environmental sustainability. These strategies are described in the UN Millennium Project's report *Investing in Development: A Practical Plan to Achieve the Millennium Development Goals,* which was coauthored by the coordinators of the UN Millennium Project task forces.

In *Prescription for Healthy Development: Increasing Access to Medicines,* the Working Group on Access to Essential Medicines of the Task Force on AIDS, Malaria, TB, and Access to Essential Medicines underscores the vital need to increase the availability, affordability, and appropriate use of medicines in developing countries. The working group proposes concrete and practical steps to increase incentives for research for priority diseases of developing countries, improve procurement and distribution, strengthen primary health systems, develop more human resources, and increase health funding. These are all necessary components of a comprehensive strategy to improve access to essential medicine in developing countries.

The working group benefited from the contributions of experts from academia and nongovernmental organizations, practitioners in the field, and members of the pharmaceutical industry. This diverse and accomplished group was able to reach consensus on most of the substantive recommendations, but ultimately, because of differences in perspective in a few areas, the representatives of the research-based pharmaceutical industry decided to withdraw their support for the report. In an appendix to the report, industry representatives explain these specific points of contention.

I am grateful for the thorough and skilled efforts of the entire working group. The practical options for action in this report should make an important contribution to achieving all the Millennium Development Goals. I strongly recommend it as required reading for anyone interested in how the world can ensure access to essential medicines in developing countries.

Jeffrey D. Sachs
New York
January 17, 2005

Contents

Boxes

Figures

Map

Tables

Working group members

Members of the Working Group on Access to Essential Medicines

Paula Munderi, Coordinator of the Working Group on Access to Essential Medicines, Medical Research Council Programme on AIDS, Uganda Virus Research Institute, Entebbe, Uganda

K. Balasubramaniam, Advisor and Coordinator, Health Action International Asia-Pacific, Colombo, Sri Lanka

Henk den Besten, Director, International Dispensary Association, Amsterdam, the Netherlands

Andrew L. Creese, Health Economist, Policy, Access and Rational Use, Essential Drugs and Medicines Policy Department, World Health Organization, Geneva, Switzerland

Mary Hlalele, Integrated Management of Childhood Illness and Reproductive Health Advisor, Management Sciences for Health, Pretoria, South Africa

Hans V. Hogerzeil, Director, Department of Medicines Policy and Standards, World Health Organization, Geneva

Bill Haddad, Chairman/Chief Executive Officer, Biogenerics, Inc., Patterson, N.Y., United States

Ellen 't Hoen, Access to Essential Medicines Campaign, Médecins Sans Frontières, Paris, France

Dan C.O. Kaseje, Director, The Tropical Institute of Community Health and Development (TICH), Kisumu, Kenya

Donna Asiimwe Kusemererwa, Assistant Manager, Joint Medical Store, Kampala, Uganda

Richard Laing, School of Public Health, Center for International Health Boston University, Boston, Mass., United States (now at World Health Organization, Geneva)

Beryl Leach, Coordinator (through April 2004), Health Action International Africa, Nairobi, Kenya

David Lee, Deputy Director, Technical Strategy and Quality Assurance, Center for Pharmaceutical Management, Management Sciences for Health, Arlington, Va., United States

Yuanli Liu, Assistant Professor of International Health, Department of Population and International Health, Harvard School of Public Health, Cambridge, Mass., United States

Zafar Mirza, former National Executive Coordinator, The Network for Consumer Protection, Islamabad, Pakistan

Eva Ombaka, Coordinator, Ecumenical Pharmaceutical Network, Nairobi, Kenya

Jonathan D. Quick (task force member 2002–03), former Director, Department of Essential Drugs and Medicines, World Health Organization, Geneva, Switzerland

Eloan dos Santos Pinheiro, former Executive Director, Fundacão Oswaldo Cruz/Far-Manguinhos, Rio de Janeiro, Brazil

Sri Suryawati, Head, Department of Clinical Pharmacology, and Director, Center for Clinical Pharmacology and Drug Policy Studies/World Health Organization Collaborating Center for Research and Training on Rational Drug Use, Gadjah Mada University, Yogyakarta, Indonesia

Task force associate

Joan E. Paluzzi, Partners In Health, Boston, Mass., United States

Preface

It is estimated that between 1.7 and 2 billion people worldwide have inadequate or no access to life-saving essential medicines. The vast majority of these people live in developing countries. After the presence of trained health professionals, medicines are the single most critical element in the maintenance of health and the successful treatment of disease and illness. Shortages of essential medicines undermine the ability of healthcare workers to respond appropriately to patient needs and this in turn often erodes the confidence and trust patients and their families have in local health systems.

The sequence of steps required from the conceptualization and production of medicines to the dispensing of them are numerous and, at times, complex. Medicines, seen as marketable commodities by many, are subject to trade and commerce policies and regulations on both national and international levels.

Underlying the specific constraints to access to medicines are the social and cultural conventions that can disproportionately prevent women, children, ethnic minorities, and other marginalized populations from gaining access not just to medicines but to the larger health system.

The Working Group on Access to Medicines of the Task Force on HIV/AIDS, Malaria, TB, and Access to Essential Medicines is composed of respected individuals who bring an impressive range of public and private experience in the many and complex local, national, and international issues that, along with availability of human and fiscal resources in any given setting, will ultimately determine access to essential medicines.

Acknowledgments

As with any large group effort, the final report remains the product and responsibility of the entire group. However, it is important to acknowledge the roles and contributions of individuals. Working group members would like to thank Alec Irwin and Eva Ombaka for the tremendously helpful background paper that launched the project and effectively got members thinking and contributing from the first meeting. William Haddad offered valuable contributions to ensure the voice of the generics industry had full hearing in the process. Ellen 't Hoen provided important information in many areas of this report; the extensive and all-important on-the-ground experience and expertise of Médecins Sans Frontières informed the report in valuable ways.

A special thanks is due Graham Dukes, who served as the main writer and editor of the interim report, much of which remained the basis of the final report. A sincere thanks is owed to Hans Hogerzeil and Margaret Kruk, who worked diligently under very challenging circumstances to help compile the final report. Detailed written reviews of the penultimate draft by Henk den Besten, David Lee, Yuanli Liu, and Eva Ombaka were extremely helpful.

We are grateful to the individuals who served as external reviewers and, who, within a fairly constrained period of time, provided us with their insightful comments and suggestions: Professor Ranjit Roy Chaudhury, President, The Delhi Society for the Promotion of Rational Drug Use, Mr. Andy Gray of the Nelson R. Mandela School of Medicine, South Africa, and Dr. Harvey Bale, Director General of the International Federation of Pharmaceutical Manufacturers Associations.

We wish to thank Professor Jeanne Bisilliat for her overview of gender and access issues during our 2003 meeting in Geneva, which informed our thinking and analysis in important ways. We also extend our thanks to Dr. Joseph Serutoke, Essential Drugs Advisor, World Health Organization Country Office

Uganda, for his inputs during our June 2004 meeting in Kampala and for his detailed, patient, and informed responses to our many inquiries throughout the development of the Uganda case study found in appendix 1. We are grateful to Kevin Burns who conducted in-depth research for and contributed to the development of the Uganda case study during his internship at Partners In Health.

Partners In Health served as the secretariat for the task force since the onset of the Project, and we thank them for their helpful and efficient logistical and administrative support.

We give our heartfelt thanks to the tireless and expert staff in the UN Millennium Project Secretariat managed by John McArthur at UN Headquarters, including Michael Faye, Michael Krouse, and Alice Wiemers. The Secretariat gave our working group its total support, and our report has benefited greatly from this commitment.

Abbreviations

AIDS	acquired immunodeficiency syndrome
CMH	Commission on Macroeconomics and Health
DFID	Department for International Development (UK)
DOTS	recommended control strategy for tuberculosis
DNDi	Drugs for Neglected Diseases initiative
GATB	Global Alliance for TB Drug Development
GAVI	Global Alliance for Vaccines and Immunizations
GDF	Global Drug Facility
GDP	gross domestic product
GFATM	Global Fund to Fight AIDS, Tuberculosis, and Malaria
GMP	Good Manufacturing Practice (WHO requirements)
HAI	Health Action International
HIV	human immunodeficiency virus
IAVI	International AIDS Vaccine Initiative
ICH	International Conference on Harmonization (of Technical Requirements for Registration of Pharmaceuticals for Human Use)
IMF	International Monetary Fund
MAP	Multi-Country AIDS Program (World Bank)
MDR-TB	multidrug-resistant tuberculosis
MMV	Medicines for Malaria Venture
MSF	Médecins Sans Frontières
MSH	Management Sciences for Health
NGO	nongovernmental organization
PEPFAR	President's Emergency Plan for AIDS Relief (US)
PRSP	Poverty Reduction Strategy Paper
R&D	research and development
TB	tuberculosis

TDR	Programme for Research and Training in Tropical Diseases
TRIPS	Trade-Related Aspects of Intellectual Property Rights agreement
UNAIDS	Joint United Nations Programme on HIV/AIDS
UNCTAD	United Nations Conference on Trade and Development
UNDP	United Nations Development Program
UNFPA	United Nations Population Fund
UNICEF	United Nations Children's Fund
UNIDO	United Nations Industrial Development Organization
USAID	United States Agency for International Development
VAT	value-added tax
WHO	World Health Organization
WTO	World Trade Organization

goals

Millennium Development Goals

Goal 1

Eradicate extreme poverty and hunger

Target 1.
Halve, between 1990 and 2015, the proportion of people whose income is less than $1 a day

Target 2.
Halve, between 1990 and 2015, the proportion of people who suffer from hunger

Goal 2

Achieve universal primary education

Target 3.
Ensure that, by 2015, children everywhere, boys and girls alike, will be able to complete a full course of primary schooling

Goal 3

Promote gender equality and empower women

Target 4.
Eliminate gender disparity in primary and secondary education, preferably by 2005, and in all levels of education no later than 2015

Goal 4

Reduce child mortality

Target 5.
Reduce by two-thirds, between 1990 and 2015, the under-five mortality rate

Goal 5

Improve maternal health

Target 6.
Reduce by three-quarters, between 1990 and 2015, the maternal mortality ratio

Goal 6

Combat HIV/AIDS, malaria, and other diseases

Target 7.
Have halted by 2015 and begun to reverse the spread of HIV/AIDS

Target 8.
Have halted by 2015 and begun to reverse the incidence of malaria and other major diseases

Goal 7

Ensure environmental sustainability

Target 9.

Integrate the principles of sustainable development into country policies and programs and reverse the loss of environmental resources

Target 10.

Halve, by 2015, the proportion of people without sustainable access to safe drinking water and basic sanitation

Target 11.

Have achieved by 2020 a significant improvement in the lives of at least 100 million slum dwellers

Goal 8

Develop a global partnership for development

Target 12.

Develop further an open, rule-based, predictable, nondiscriminatory trading and financial system (includes a commitment to good governance, development, and poverty reduction—both nationally and internationally)

Target 13.

Address the special needs of the Least Developed Countries (includes tariff- and quota-free access for Least Developed Countries' exports, enhanced program of debt relief for heavily indebted poor countries [HIPCs] and cancellation of official bilateral debt, and more generous official development assistance for countries committed to poverty reduction)

Target 14.

Address the special needs of landlocked developing countries and small island developing states (through the Program of Action for the Sustainable Development of Small Island Developing States and 22nd General Assembly provisions)

Target 15.

Deal comprehensively with the debt problems of developing countries through national and international measures in order to make debt sustainable in the long term

Some of the indicators are monitored separately for the least developed countries, Africa, landlocked developing countries, and small island developing states

Target 16.

In cooperation with developing countries, develop and implement strategies for decent and productive work for youth

Target 17.

In cooperation with pharmaceutical companies, provide access to affordable essential drugs in developing countries

Target 18.

In cooperation with the private sector, make available the benefits of new technologies, especially information and communications technologies

Executive summary

Medicines are the most significant tool that society possesses to prevent, alleviate, and cure disease.[1] Most illnesses, especially infectious diseases, are either preventable or to some extent treatable with a relatively small number of existing medicines. Combined with appropriate public health interventions, appropriately prescribed essential medicines and vaccines could, in principle, massively reduce the impact of disease on communities, especially children (WHO 2004a).

The problem

A very large part of the world's population has inadequate or no access to essential and life-saving medicines. According to one study, more than 10 million children die unnecessarily each year, almost all in low-income or poor areas of middle-income countries, mostly from a short list of preventable diseases such as acute respiratory diseases, diarrhea, malaria, measles, and causes related to malnutrition (Black 2003).

The lack of access to life-saving and health-supporting medicines for an estimated 2 billion poor people stands as a direct contradiction to the fundamental principle of health as a human right. Poverty and illness create a vicious cycle. Poverty is at the source of major health risks, such as insufficient and improper nutrition, poor sanitation and hygiene, toxic indoor smoke, and extremely limited access to health education and services, all of which determine almost 45 percent of the disease burden in Least Developed Countries (WHO 2002g). Illness is a major reason that the nearly poor slip into poverty. Illness decreases people's ability to work (be it remunerative or nonremunerative). Illness prevents children from getting the education they need. Women and children make up the majority of the poor, and their low status in many societies often means that they have even less access to medicines. Improving access to medicines must be a key component of strategies to strengthen healthcare.

Increasing access to medicines, especially for the poor, offers many challenges

The knowledge and medicines are available to reduce the incidence of death and suffering greatly; what is still needed is clear priority setting and the provision of adequate resources. Resolving some of the greatest health crises in human history will not be the sole responsibility of any single actor or sector of society. These challenges come at a time when an unprecedented number of the world's population is also living (and dying) in extreme poverty. Public, private, and nongovernmental organizations (NGOs) and institutions must work together.

Increasing access to medicines in developing countries, especially for the poor, offers many challenges. These can be crystallized into two main areas:

- How to increase access to affordable existing medicines in resource-poor settings, which countries can do by improving the selection and use of essential medicines, taking steps to ensure affordable prices, increasing sustainable financing, and strengthening reliable supply systems.
- How to find new ways to promote the development of new medicines and vaccines to treat diseases of poverty.

The World Health Organization (WHO) estimates that nearly a third of the world's population lacks access to the most basic essential medicines, while in the poorest parts of Africa and Asia this figure climbs to a half (WHO 2000a; WHO 2003f). WHO has also estimated that in Africa and Southeast Asia, prompt diagnosis and treatment with appropriate medicines could save approximately 4 million lives annually (DFID 2004b). Moreover, it is often the poorest people who are paying the highest out-of-pocket expenses for medicines because the public sector in developing countries fails to provide affordable medicines reliably. Medical insurance schemes cover less than 8 percent of the population in Africa, and these schemes may not cover prescription medicines on an outpatient basis. Participatory assessments during national poverty reduction strategy processes often elicit the availability of medicines as a primary indicator of the effectiveness of healthcare delivery.

Though access to essential medicines has improved in recent years, WHO reports that delivering "the right medicines to the people who need them at the time they need them remains a major challenge" (WHO 2004a, p. 61).

Figure 1
Distribution by country income group of people without access to essential medicines, 1999

Source: WHO 2004a.

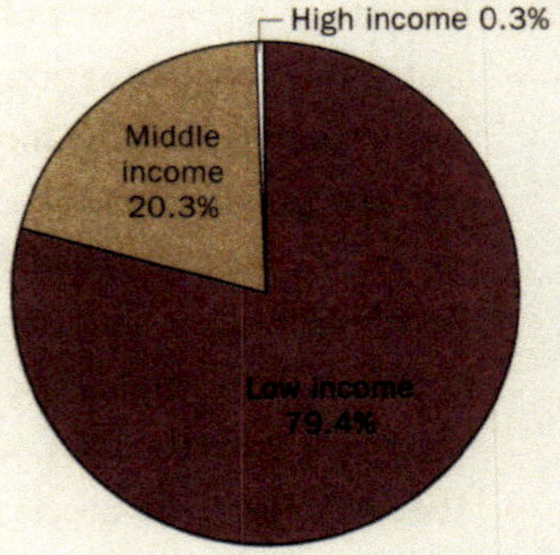

Increasing access to affordable medicines requires improving selection, prices, and supply

The analysis contained in WHO's 1999 *World Medicines Situation* shows that roughly two-thirds of the world's population now have regular access to essential medicines, up significantly from 1975 when this proportion was just under one-half (WHO 2004a). However, global population growth has meant that the absolute number of people without access has remained nearly constant, at approximately 1.7 billion.

Despite the progress made in the last decades, the likelihood of an individual having access to essential medicines is still greatly affected by income level. The *World Medicines Situation* analysis found that people in poorer countries were much less likely to have access to these medicines (WHO 2004a). According to WHO, in 1999, roughly 80 percent of the global population without access to essential medicines was living in low-income countries (figure 1). This is a disproportionate share of the global burden, given their estimates that low-income countries account for approximately 60 percent of the world's population (WHO 2004a). In contrast, only 0.3 percent of those lacking access to essential medicines lived in high-income countries, which account collectively for about 15 percent of the world's population. In a global context, 15 percent of the world's population consumes 91 percent of the medicines produced (WHO 2000a). Of people living in low-income countries, nearly 40 percent did not have access to essential medicines in 1999 (WHO 2004a).

Geographically, the lack of access to essential medicines is especially severe and concentrated in Africa and India (figure 2). In fact, 38 percent of the people without access to essential medicines live in India, and 15 percent live in Africa (WHO 2004a). Together, India and Africa account for 53 percent of the world's population without access to essential medicines (WHO 2004a). Although the disease burden and mortality from preventable or curable illness is highest in Africa, pervasive poverty means that the continent's share of the global pharmaceutical market is only slightly more than 1 percent.

India's and Africa's inordinate share of the global population without access is not entirely a function of population. India accounts for 17 percent of the world's population, while Africa accounts for roughly 10 percent of the world's population. This translates to very high absolute numbers of people without

Figure 2
Distribution by region of people without access to essential medicines, 1999

Source: WHO 2004a.

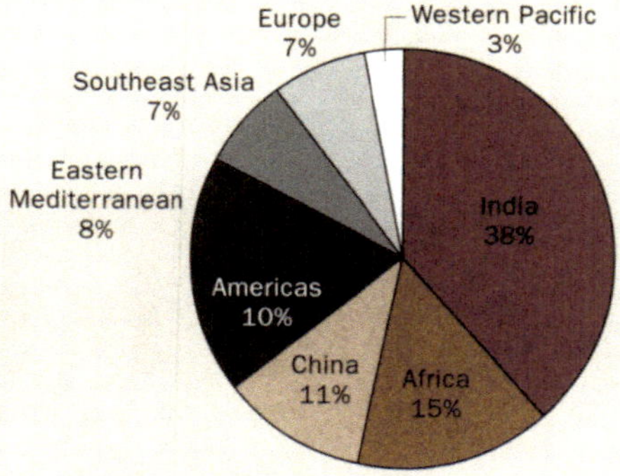

Increasing access also requires promoting the development of new medicines and vaccines

access in these two regions. Sixty-five percent of Indians and 47 percent of Africans lack access to essential medicines (figure 3). In Europe that share is 14 percent and in the Americas 22 percent (WHO 2004a).

The lack of access to medicines in most developing countries reflects the lack of sufficient incentives for developing new medicines to target those communicable diseases that disproportionately afflict the poorest countries, and also their inability to pay for and effectively distribute those that do exist. The result is what the government of the United Kingdom has called a "mismatch between pharmaceutical needs in developing countries and the current nature of the global pharmaceutical market" (DFID 2004a, p. 14).

When examining access to essential medicines for the poor, the Working Group on Access to Essential Medicines identified a number of fundamental problems common to many countries. What is very clear, however, is that the basic knowledge and technical information already exist to increase access to all segments of a population. Furthermore, the world possesses the resources to fund adequate access to essential medicines and functioning health systems in the developing world.

Ideally, the working group would have wished to propose a simple approach to improving access to medicines. Because obstacles to access are many and diverse, and because they differ in nature and degree from one country to another, this is not possible. The issues surrounding access are complex, at times culturally specific, and often fluid; the solutions can be no less. Some will need to be applied at the global level, while others need to be selectively employed in particular countries or regions depending on the situation and its context.

Overarching barriers

At the risk of overshadowing the other vital issues presented in this report, the Working Group on Access to Essential Medicines points to six of the most

Figure 3
Share of people without access to essential medicines by region, 1999

Source: WHO 2004a.

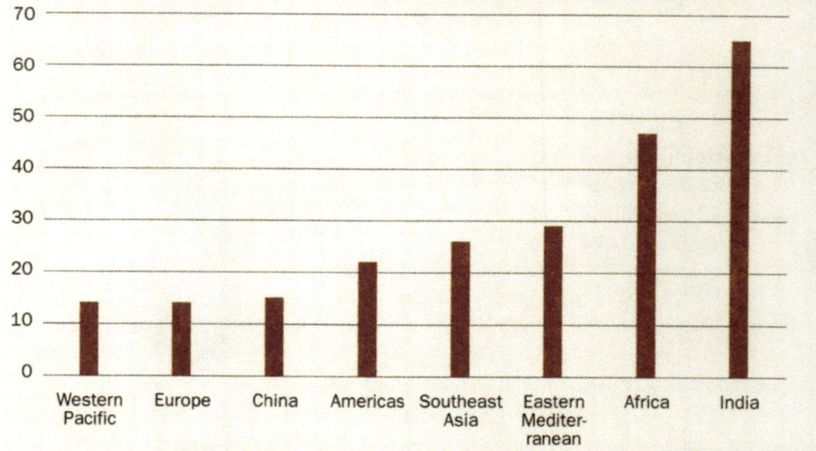

Inadequate human resources threatens to undermine all efforts to strengthen health systems and improve healthcare

important barriers to access that merit special attention and action at this time. The first four relate to barriers to existing medicines; the last two constitute barriers to the development of affordable and available new medicines and vaccines.

Barriers to existing medicines

1. *Inadequate national commitment* to making healthcare a priority from the national to the local levels remains one of the greatest barriers to increasing access to existing medicines. There are many reasons for this lack of prioritization. Key among them are a lack of political will by policymakers to make the needs of the poor a priority; donor programs that can skew or limit national governments' abilities to set health policy; debt servicing and conditionality for loans from international financial institutions that can further limit government responsiveness to basic social service needs of citizens; and, unfortunately, the threat of corruption that continues in the healthcare sector at all levels.
2. *Inadequate human resources* for health, including pharmacists and pharmacy technicians, is a growing problem that, if unaddressed, threatens to undermine all efforts to strengthen health systems and improve healthcare in much of the developing world. Education, information, and in-service training remain potent tools to change that situation. More needs to be done to identify what is needed to retain skilled workers, especially in the face of mounting demands for health workers, such as nurses and pharmacists, in developed countries. Retention plans and compensation schemes for countries that lose health workers should be investigated.
3. The *international community has not provided adequate finance nor consistently fulfilled its existing promises to developing countries.* Some proposed actions have not been carried out at all and others have not been carried out effectively. To achieve progress, there will be a need for political will, in both industrialized and developing countries, as well as a need for transparency on all fronts. Above all, there will be a need for increased levels of long-term financial support from the world community. It remains an unfortunate ongoing reality that some of the world's wealthiest countries remain the farthest from achieving their longstanding commitment to the international development assistance target of 0.7 percent of gross domestic product (GDP).
4. A persistent *lack of coordination of international aid* reduces access to medicines. Most poor countries will require significant donor funding to achieve universal access to essential medicines. They will also need much better aid coordination to avoid unnecessarily heavy reporting requirements and to avoid resource-wasting duplication of efforts. Sector-wide approaches should be used to promote improved coordination.

A need exists for a great deal more transparency and coordination of effort

Donors should commit aid that strengthens existing systems, that proactively targets the poorest and rural areas, and that avoids vertical programming by disease or by a given donor. A need exists at both the international and national levels for a great deal more transparency and coordination of effort between the large number of organizations that have already become involved in one way or another in this field. The involvement of so many bodies can and does lead to duplication of effort and to waste, and both are unacceptable. In some situations, there is every reason to merge complementary ventures. Pharmaceutical companies can and should contribute in their own way to the advancement of national medicines policies and the development of capacity in this field.

Barriers to the development of affordable new medicines

5. *The Trade-Related Aspects of Intellectual Property Rights (TRIPS) agreement may block access to affordable new medicines and vaccines.* After January 2005, generic production in India, the source of many vital existing medicines for developing countries without productive capabilities, will be fully subject to TRIPS provisions (WTO 1994). Concerns also exist that the August 30, 2003, decision reached by the WTO General Council concerning a waiver for TRIPS Article 31(f) (which would allow a compulsory license to be issued by the country in need and by the country that can produce the medicine for export) will be too cumbersome for developing countries to exploit (WTO 2003). Finally, the growing number of bilateral and regional trade agreements with major trading partners, such as the United States and the European Union, may often contain provisions that limit developing countries' use of existing flexibilities under TRIPS to protect public health (such as restrictive compulsory licensing conditions and parallel importation provisions, extended data protection, and requiring medicines regulatory agencies to take on national patent office oversight duties).

   A major recommendation of this working group is for the WHO to be mandated, perhaps in coordination with the WTO or other trade bodies, to monitor the impact of TRIPS compliance by major developing country exporters and, in particular, to monitor the use of the August 30 decision over the coming two years as it concerns access to medicines. A report to the World Health Assembly and the WTO General Council, with recommendations, should be delivered no later than the end of 2007.
6. *The current incentive structure is inadequate to promote research and development of medicines and vaccines to address priority health problems of developing countries.* For a number of the most neglected diseases (such as African trypanosomiasis, Chagas disease, leishmaniasis, and dengue fever), which

Every developing country should have an overall national medicines policy and strategy founded on the essential medicines concept

occur primarily in developing countries, new medicines need to be developed (WHO and IFPMA 2001). For others, new medicines are needed to address shortcomings of existing treatments, such as safety, efficacy, appropriate dosing, length of treatment, and the ongoing threat of drug resistance. Despite progress in funding research and development (R&D) for new medicines for neglected diseases, with notable contributions from philanthropic foundations and some governments and pharmaceutical companies, more financial resources need to be mobilized in a sustainable way to create a strong pipeline of new medicines. New thinking, different means of financing and organizing medicines development, and other reforms are needed. For example, the WHO Commission on Intellectual Property Rights, Innovation, and Public Health should examine alternative international models to the current patent-based system for priority setting and financing of health R&D.

Since the problem of access has many causes, a single solution to improve the provision of medicines cannot be expected to succeed alone; it must be complemented by others. What this means is that every developing country should have an overall national medicines policy and strategy founded on the essential medicines concept:

> A national drug policy is a commitment to a goal and a guide for action. It expresses and prioritizes the medium- to long-term goals set by the government for the pharmaceutical sector, and identifies the main strategies for attaining them. It provides a framework within which the activities of the pharmaceutical sector can be coordinated. It covers both the public and the private sectors, and involves all the main actors in the pharmaceutical field. (WHO 2003a, p. 19)

The most important step is to develop a model that meets national needs and to employ it as a basis for developing and managing the system as a whole.

National medicines policy cannot succeed in isolation from broader health policies and government policies in general. A ministry of health is unlikely to succeed in this area unless it has clear and acceptable understandings reached with other government departments dealing with such matters as finance; the training curricula for health professionals; the salaries of public employees; and practices regarding trade, taxation, and customs duties, all of which are likely to impinge on the supply of medicines.

Health sector strengthening and development to reach the Millennium Development Goals should be done in coordination with national poverty reduction strategic planning being adopted in poor, indebted countries. A recent WHO review of some national Poverty Reduction Strategy Papers (PRSPs) and health pointed out that much good information is being generated about health system needs, including access to medicines, in analytical phases of the poverty reduction strategy process (WHO 2004e). However,

Achieving universal access to affordable essential medicines requires developing and strengthening primary health systems

it would appear that when government responses were formulated, there was a tendency to rely on existing health policy approaches and budgets, which rarely included any community or civil society participatory processes. Instead, they tended to reflect top-down prioritization shaped by international financial institution conditions and a lack of political commitment to reorient government focus more toward the social sector. A need exists, therefore, to examine how health sector and other social sector needs can be better addressed in the poverty reduction processes, including how information and needs assessments gained in participatory analysis can be better translated into government planning and budgets.

Reaching the goal of achieving universal access to affordable essential medicines in developing countries will require the development and strengthening of primary health systems, along with the myriad specialized administrative and functional features necessary to maintain a sustainable supply of all essential medicines. Firm priorities must be set, including an urgent need to reassess the importance of health sector investments by donors and recipient governments and to increase resources substantially through greater political will on both sides.

Summary of main recommendations

The Working Group on Access to Essential Medicines has organized its analysis and recommendations into three main categories: availability, affordability, and appropriateness. The group also identified basic principles and crosscutting issues.[2]

Expanding access to essential medicines requires attention to a diverse set of policy challenges. National health policies and systems are not always fully attuned to ensuring that medicines are available, affordable, or appropriate. Solutions must begin with an understanding of local health conditions in their broadest epidemiological, economic, regulatory, and even cultural contexts. Increasing access must be seen as a process requiring ongoing support from a range of stakeholders. Reforms are most effective when they focus on the most critical access problems, rather than attempting to address all barriers simultaneously. Countries need adequate data collection and analysis to assess and set priorities in problem areas.

Access to medicines cannot be addressed in isolation either from the rest of the health system or from the overall health situation in a given country. Access to medicines is not an issue that exists in a vacuum: it is an integral part of healthcare, the various components of which are mutually supportive. Measures in all of these areas will need to be backed by the systematic and ongoing assessment of the needs of a particular country or population. On all levels there will be a need for institutional development and a sustainable expansion of human resources. Although the access to medicines issue ultimately is a global one, the working group, in keeping with the UN Millennium Project's

Women's inequality and gender disparities contribute to institutionalized inequalities within education and health systems

mandate, has focused its main efforts on addressing how to increase access to medicines in developing countries that have the greatest need for concerted, coordinated, and effective mobilization of resources to break the cycle of ill health, poverty, and declining economies.

The working group addressed its recommendations on two main levels: national and international. Especially at the national level, an attempt was made to be as operational as possible. This division into national and international levels, however, should be viewed with caution, since increasing access will ultimately involve a complex interplay of many actors operating at many levels concurrently and dynamically.

General principles

The working group found that certain basic principles underpinned approaches to the issues of and solutions to increasing access to medicines. These general principles include the human right to health codified in the UN Declaration of Human Rights (UN 1948); the right to treatment codified in Article 12 of the International Covenant on Economic, Social, and Cultural Rights, which was clarified in 2000 to include the right to essential medicines (WHO 2002a; Hogerzeil 2003); and the right to medical treatment, including access to medicines, found in the International Guidelines on HIV/AIDS and Human Rights (OHCHR/UNAIDS 2002). However, the enforcement of these rights is not evident in the current global situation, where entire populations, particularly the poor and underprivileged, commonly have little or no access to essential medicines.

The working group also found that women's inequality and gender disparities contribute to institutionalized inequalities within education and health systems. These inequalities limit women's and girls' access to healthcare and to needed medicines more than men's and boys'. Profound, incremental, and societywide changes must occur to eliminate these forms of discrimination. Health systems will need to be strengthened to deliver quality essential services while maintaining equity of access. Equity of access should be a cornerstone in thinking and policymaking. Simply put, the most marginalized should receive healthcare and services at the same or greater rate than more economically franchised in any program. In the case of AIDS treatment, an equitable approach would target populations that live in the most resource-challenged areas first. An equitable approach to pro-poor healthcare, would be, by definition, a bottom-up approach.

The working group in general also recognized the need to find new ways for the main actors involved in the supply of pharmaceuticals to interact to ensure that needed medicines are available. Indeed the group's discourse on the means to ensure supply was vibrant and robust. Members did agree on the fundamental point that market competition is an essential driver for innovation, supply, and affordable prices.

Medicines research must become better attuned to the needs of the world as a whole

Improving the availability of medicines

Availability of medicines is affected by many factors. The main ones that need to be tackled include ensuring that needed medicines are developed and brought to market, and that supply and distribution systems are adequate to deliver them to the people who need them.

Improving the rate and relevance of innovation. Treating priority diseases of the poor is greatly hindered by a fundamental problem: the medicines required for some of the diseases and illnesses most prevalent in developing countries do not exist because of a lack of therapeutic innovation (MSF 2001). Another critical need is for new medicines to supplement or replace those to which microorganisms have become resistant, as is notably the case for malaria and tuberculosis.

A reorientation of medicines research, better attuned to the needs of the poor, is necessary. This will require creative new research, development, and financing mechanisms. The for-profit private sector is not going to take up needed innovation for major infectious diseases of poverty without major involvement and subsidy from the public sector and an appropriate and supportive policy environment. The public sector is also going to have to remain a vital force. An equitable approach globally would have these innovation costs borne primarily by the nations with the broadest shoulders: heavily industrialized countries with strong economies that are capable of sustaining relatively high prices for the medicines they require.

Public-private initiatives, such as the Medicines for Malaria Venture, the International AIDS Vaccine Initiative, and the Global Alliance for TB Drug Development, appear to be offering useful models for new medicines and vaccines development, and they should be supported. However, some questions remain about governance, adequate participation by experts from affected countries, and adequate focus on priority medicines for the poor. These aspects need to be monitored.

Successful innovation to help meet the Goals will require greater cooperation among all sectors (such as the public and private sectors, academia, foundations, and the United Nations), substantially more financing from multiple sources, clear priorities for research efforts, effective management, and technology and knowledge transfer. WHO should take a leading role in promoting R&D that meets the public health priorities of developing countries. Medicines regulatory process reforms and harmonization need to better reflect and serve the needs of developing countries. Traditional knowledge and medicines continue to be marginalized, to the detriment of consumers. Vigilence surrounding all aspects of pharmacological practice in developing countries needs to be strengthened.

All of these issues point to the need for considerable change, which will take considerable time to implement and to produce results. Taking new steps must start now.

International standards for ethical research should be applied in all countries

At the national level:

- Governments should determine priorities in medicinal innovation in accordance with the most basic and unfulfilled needs of their populations, and bring these priorities to the fore both in their domestic policies (such as through their national medicines policies, essential medicines lists, procurement strategies and budgets, and public R&D policy priority setting) and in the global forums in which they participate.
- Developing countries should be more confident about negotiating for technology transfer and more national capacity building to participate directly in R&D. Examples of innovative approaches include the Drugs for Neglected Diseases initiative (DNDi) approach to partnering with research institutes in developing countries, the cooperative effort between the Universities of Nairobi and Oxford on AIDS vaccines trials, the Kenya Medical Research Institute's partnering with GlaxoSmithKline and the University of Liverpool on the development of a new antimalarial, the Merck Vaccine Network Africa training center at Moi University in Kenya, Merck's partnership with Harvard University's AIDS Program in the Enhancing Care Initiative to build infrastructure for vaccine delivery, and the Pfizer partnership with Makerere University in Uganda and the University of Utah. Even in countries with very limited resources, some steps can and must be taken to formulate a national research policy and provide the funding and infrastructure needed to implement it, either independently or in collaboration with foreign, regional, or global institutions.
- The regulatory environment should reward sound research into priority diseases. For example, a country could devise a fast-track system for priority medicines, based on national health priorities.

At the international level:

- Public investment in medicinal research should be expanded to meet the most pressing needs of developing countries and poor populations, including developing knowledge on the basis of indigenous medicines. The international community should not rely on the research-based pharmaceutical industry to be the primary vehicle for developing medicines needed in developing regions. New ways of approaching innovation should be considered and pursued with some urgency (Folb 2004). The WHO Commission on Intellectual Property Rights, Innovation, and Public Health should examine alternative international models to the current patent-based system for priority setting and financing of health R&D. Recent papers commissioned by DFID also support the value of taking new approaches to technology transfer, patent regimes, intellectual property management, and local production as ways of meeting the demand for increased access to medicines (see, for example, Hill and Johnson 2004; Lewis-Lettington and Banda 2004).

National regulatory bodies need strengthening, and the judicial system needs the human and material resources to enforce regulations and eliminate abuses

- International standards for ethical research such as those elaborated by the Declaration of Helsinki should be applied in all countries.

Developing more reliable procurement and supply systems. Many national procurement and supply systems for medicines, whether public or private, are inefficient or poorly attuned to current needs. Procurement is not always in line with what is needed, funds are not optimally used, and medicines are commonly out of stock in both urban and rural areas. Procurement and supply systems in developing countries need to be more effective and reliable, making the best possible use of public, private, and nonprofit channels and ensuring that a reliable supply system is extended to rural areas. Each country should develop and keep an updated list of essential medicines that reflects its priority health needs and that is used as a basis for procurement and supply decisions.

At the national level:

- All potentially efficient systems for the procurement and distribution of supplies of medicines, whether public, private, or maintained by NGOs, should be encouraged and assisted to develop. This will require country-level, ongoing capacity building. Low-income countries especially need ongoing technical assistance to build expertise in effective procurement, quality control, and quality assurance systems. National regulatory bodies need strengthening urgently in developing countries, and the judicial system should be provided with the human and material resources to enforce these regulations and eliminate abuses that can lead to waste and loss.
- The advent of the Global Fund to Fight AIDS, Tuberculosis, and Malaria (GFATM) in recent years and the World Bank Multi-Country HIV/AIDS Program (MAP) provide developing countries with valuable resources and incentives to improve their procurement and medicines management systems. Both organizations promote an assured quality and lowest price approach. The GFATM asks recipient countries that receive funds for medicines purchases to demonstrate that they have a competent national system for selection, procurement, quality assurance, supply, and distribution. Initial concerns that the fund would prompt parallel procurement and supply systems are being allayed. The fund has emphasized that national systems should be strengthened, not replaced nor sidelined. The World Bank published a detailed technical guide in early 2004 that should be very helpful at the country level in addressing these systems challenges (World Bank 2004a).
- The WHO prequalification project to identify good-quality products for HIV/AIDS, tuberculosis, and malaria medicines for procurement by UN agencies is also helping low-income countries that have very limited quality assurance capacities to improve procurement by providing key quality indicators for suppliers and products. The working

Technical assistance should strengthen national systems to be able to protect and promote public health for all

group endorses the WHO prequalification project for use by developing countries and supports its expansion.

- Pooled procurement schemes remain a tantalizing, yet still underused, avenue for improved procurement. No one model for pooled procurement exists. The degree of cooperation and shared or combined systems depend on the participants, local and regional characteristics, and purchasing needs. Other examples of pooled procurements include disease-specific international initiatives such as the Global Drug Facility for TB and the Medicines for Malaria Venture. All of these strategies should be explored by developing countries. Countries pursuing these strategies should take care that a minimum number of qualified suppliers participate in these schemes to ensure a competitive market.
- Procurement should be only from suppliers that have complied with the WHO Good Manufacturing Practice (GMP) requirements.

At the international level:

- The exchange of information and advice on successes and failures of national or pooled procurement systems, routinely updated price lists, and systems of distribution and supply will be valuable in establishing new agencies or reforming those that already exist. Bringing together data from many countries on current and anticipated needs and priorities will create a basis for producers to provide appropriate supplies. International standards for operating procurement agencies are needed, and ways to prequalify procurement agencies that attain these standards should be developed.
- WHO, the GFATM, and the World Bank should provide leadership in meeting capacity-building demands. WHO, the World Intellectual Property Organization, WTO, and especially competent nongovernmental experts should provide country-level guidance on the effects of intellectual property regulation on access to medicines. The goal of all technical assistance should be to strengthen national systems to be able to protect and promote public health, particularly for the poor and marginalized. Countries that do not have sufficient regulatory capacity in the short or medium term should have access to international bodies, norms, and standards to help them make efficient decisions about quality assurance, quality control, and registration.

Promoting the safety of medicines. Substandard medicines (genuine products that do not conform to the pharmacopeial standards set for them) present a real problem, especially in developing countries that have limited regulatory and enforcement capacities. Use of these medicines endangers lives, wastes scarce resources, and contributes to development of resistance to anti-infectives. WHO estimates that as many as 200,000 of the more than 1 million deaths from malaria each year could be avoided if medicines were effective, of good quality, and used correctly.

Registration procedures should be simple, straightforward, and equitably applied

WHO also reports that the US Food and Drug Administration estimates that more than 10 percent of medicines in circulation in both developed and developing countries is counterfeit (products that are deliberately and fraudulently mislabeled with respect to identity or source). A WHO survey of counterfeit medicines reports from 20 countries showed that 60 percent of counterfeits were found in poor countries and 40 percent in industrialized countries (WHO 2003d). A recent report from the United States Pharmacopeia and Drug Quality and Information programs on the quality of anti-infectives in Asia indicates that the availability of substandard and counterfeit drugs has reached a disturbing proportion in resource-poor settings (USP and DQI 2004). This report identified gaps and weaknesses:

- Weak national medicines regulatory authority and weak enforcement of relevant laws.
- Little or no GMP compliance by manufacturers.
- Limited laboratory capacity in terms of qualified staff and equipment.
- Lack of competent inspectors.
- Lack of inexpensive, quality-assured medicines (USP and DQI 2004).

At the national level:

- Countries can combat the sale and use of poor-quality medicines by raising public knowledge and empowering consumers to demand quality assurances, conducting additional inspections of companies suspected of producing or importing substandard or counterfeit drugs, strengthening medicines laws, imposing stiffer penalties for offenders, increasing postmarketing surveillance, and restructuring the regulatory system. However, governments that rely on donor funding find themselves constrained in calling for system strengthening and increased staffing, given externally imposed conditionalities that can limit social sector spending, especially on government staffing.
- National systems that monitor suspected adverse reactions to medicines need to become more effective. They should be capable of defining the overall pattern of unwanted reactions in the population (and in particularly susceptible groups) and also cases of frank injury due to medicines. Independent drug information centers should be supported as part of improving information exchange—nationally and across borders—on medicines quality and safety. These centers must include data on benefit-risk assessment of particular agents or products, regulatory decisions involving safety issues (such as the withdrawal of disproportionately risky medicines), and reliable information on poor-quality products and producers.
- Work should be undertaken to institute no-fault systems for redressing injury from medicines.
- National registration should require bioequivalency information for both originator and generic medicines, to be provided and financed by the company seeking registration.

The drug supply systems in many developing countries are seriously underfinanced

- Registration in most developing countries takes too long for reasons that are not always clear. Registration procedures should be simple, straightforward, and equitably applied. When possible, fast-track processes should be available for medicines for national priority health needs, especially those prequalified by the WHO.

At the international level:

- International agencies and donors need to make safety and quality of medicines a higher priority by supporting regulatory strengthening and the timely exchange among countries, whether importing or exporting, of information relating to the safety of medicines. They should also enforce compliance with international GMP.
- The WHO prequalification project should be strengthened, expanded, and made a permanent and well funded function of WHO.
- Recent WHO initiatives to prequalify both individual products for high-priority diseases and the factories producing these products need to be vigorously pursued and extended.
- International organizations should share information about poor-quality products and producers based on reliable and accurate data and strengthen systems for sharing information on benefit-risk assessment and regulatory decisions (such as withdrawals). International organizations should also support existing adverse event monitoring systems.
- International organizations should work to strengthen national regulatory capacity through training, capacity building, information sharing, evaluation of best practices, and sustained funding.

Increasing the affordability of medicines

The medicines supply systems in many developing countries are seriously underfinanced. It would be unrealistic to imagine that developing countries will succeed in correcting this situation on their own during the coming 15–20 years, especially in light of growing disease burdens from major pandemics. Donor support for low-income countries should therefore be designed with a 20-year horizon at the levels needed to meet national public health goals. This support should be coordinated through sectorwide approaches.

Adequate and fair financing. Financing strategies should promote health system strengthening and progress toward national self-reliance over time.

At the national level:

- There must be a progressive increase in the public sector budget for essential medicines, particularly in order to ensure improved access for the poor; this is likely to require a shift in the allocation of government resources. Political will must exist, and government allocation decisions should be made using accurate data (such as such as those from national pharmaceutical sector baseline surveys). Governments need to

Low-income countries need long-term financial support to strengthen their systems and procure medicines

understand the importance of guaranteeing financing for procurement arrangements. User fees act as an economic barrier to healthcare for the poor. They do not provide an adequate nor long-term solution to the problem and they should be phased out wherever they exist.

- Essential medicines, along with other essential health services, should be provided at no cost to the end user in developing countries. For the poorest countries, financing in the short to medium term must come from wealthy countries, which have repeatedly committed to spending 0.7 percent of their national GDP on official development assistance and, in most cases, have fallen short. Community financing, while a useful complement to government-financed healthcare, cannot be, in the short term, a viable option for sustainable financing of primary healthcare in low-income countries.
- Payments made to providers at all levels—importers, wholesalers, and retailers—should be commensurate with the degree of service which they provide, as determined by appropriate national authorities.
- The acceptance of public or private donations of medicines should strictly follow *Guidelines for Drug Donations* (WHO 1999a).

At the international level:

- The donor community needs to accept the fact that low-income countries will need substantial additional financing per capita to meet even the most basic primary healthcare (including medicines) packages. In Uganda, for example, the spending shortfall on medicines alone is stark ($1.20 allocated per person versus an estimated need of $3.50 per person). It is cruelly cynical to suggest to poor countries that they need to prioritize healthcare for sustainable social and economic development and then not deliver the financing required. Low-income countries need long-term, sustained financial support to strengthen their health systems and procure needed medicines. In many cases, they will also require debt relief.
- Health sector budgets should be privileged in programs supervised by the international financial institutions, and levels of donor assistance should be adequate to support levels of service needed to achieve the Millennium Development Goals.
- Donors should fund recurrent costs, such as salaries, in the poorest countries for the short to medium term to enable health systems to function.
- In low-income countries, loans will occasionally be justified in order to provide acute relief, but in principle, funding should be in the form of financial grants, preferably provided without ties. Where loans are made, they should be earmarked for health systems development and not for the purchase of consumables, such as medicines. The world community can also provide valuable support in acquiring, analyzing, and disseminating

If a price is set at a level that a consumer cannot afford, the medicine will not be bought and used

comparative financial data on drug supply and the flow of finance both between and within countries; this will form a valuable basis for polices designed to ensure sufficient and equitable financing. Financing should promote integrating medicines procurement and supply with wider health policies and systems. Middle-income countries should be given incentives to allocate more of their available national budget for healthcare and medicines, with some international support being an option as needed.

- Innovative new global mechanisms to promote pharmaceutical R&D for urgent health problems of the developing world should be a priority. Although the total amount required is not clear, assessing current international funding flows and existing R&D needs is urgently required to identify the magnitude of funding required.

Countering high prices. Prices matter. If a price is set at a level that a consumer cannot afford, the medicine will not be bought and used. In developing countries, the overwhelming burden of poverty means that most essential medicines are not affordable. Yet every day, poor people risk their tenuous economic security to purchase medicines. Too often, the decision is a brutal tradeoff: food, housing, and children's education or the purchase of needed medicines. Sometimes drugs are unavoidably costly, but in a great many instances they are disproportionately expensive. The reasons for high prices are multiple, and the problem therefore has to be tackled vigorously at various levels. Market competition remains the most potent way to affect and lower prices. Additionally, the presence of an effective and efficient procurement and distribution system cannot be overemphasized.

At the national level:

- Governments have a range of tools available to help manage and lower medicines prices: use available and impartial price information; have and use an updated essential medicines list; have a pro-generics approach in policy (including mandatory substitution), planning, and procurement; promote price competition in the local market; promote bulk or pooled procurement (while taking care to maintain adequate numbers of qualified suppliers to supply the market); negotiate equitable prices for patented essential medicines; eliminate taxes (such as the value-added tax), duties, and tariffs on essential medicines; minimize mark-up; encourage local production of essential medicines where feasible; and ensure TRIPS public health safeguards are in national legislation and the expertise and will exist to use them.
- Prices for medicines should be transparent because information asymmetries are a main source of procurement inefficiencies that can result in higher prices. Medicines price lists, such as those published by WHO and Management Sciences for Health, can be a valuable tool for countries.

Strategies are needed to permit continued production and supply of low-cost generic drugs for poor populations

- Prices of medicines in developing countries must be reduced to the minimum sustainable level, which in many developing countries means that industry needs to provide these medicines at production cost ("no profit, no loss") to national health systems. In middle-income countries, differential pricing should be pursued, although the prices will not be at marginal cost.
- Governments should recognize that guarantees of timely payment and financial credibility with suppliers are extremely effective in lowering prices. Suppliers, above all, want to know that they will be paid, and that it will be in a timely manner.

At the international level:

- There is the need to identify and adopt strategies that will permit continued production and supply of low-cost generic medicines for poor populations after January 2005. This is likely to involve providing new options, beyond those already incorporated in TRIPS. Of key concern will be the impact of TRIPS compliance by India, a major source of low-cost generic essential medicines in developing countries, and overall use of the WTO August 30 decision, which may prove too cumbersome to be considered a real solution (see figures 2.1 and 2.2 for details of how Least Developed Countries and developing countries can use it). Regional and bilateral trade agreements should not compromise the ability of developing countries to invoke the flexibilities provided in TRIPS (see, for example, Vivas Eugui 2003). The impact of TRIPS compliance and the August 30 decision on access to medicines in developing countries should be monitored by competent authorities, such as WHO, and findings to date and recommendations should be reported by the end of 2007.
- Pharmaceutical companies should be willing to negotiate medicines prices based on a concept of equity.[3] Differential pricing negotiations should be simplified and transparency should be assured.
- All efforts must be made to continue and strengthen best price, assured quality procurement policies in the GFATM and the World Bank MAPs. Bilateral programs that restrict procurement only to originator medicines limit the impact of such aid to populations in great need, and such restrictions should be avoided.
- Both within exporting states and in international consultation, policies should favor international competition in the pharmaceutical field, including unhampered competition between individual firms and between originator companies on the one hand and generics producers on the other.
- Much benefit will be gained by sharing information between countries and agencies on producer prices, mark-ups and profits, tariffs, taxes, and other charges, so that successful approaches to the reduction of consumer prices in one country can be emulated in others.

Consumers often judge the quality of care by whether they received a prescription

Promoting the appropriate use of medicines

Medicines are not fully available to a population unless the treatment in which they are used is provided in such a way that the patient is most likely to benefit. In many situations, inappropriate prescribing, dispensing, and consumption of medicines means that this aim is not achieved.

Better prescribing and dispensing. In too many cases, prescribers write too many prescriptions, and they do so for many reasons. For example, quick and affordable testing for an acute respiratory infection may not be available, so a clinician will presume the worst and prescribe an antibiotic—just in case. Consumers often judge the quality of care by whether or not they received a prescription. In some developing countries, the average number of prescriptions per visit can exceed three. Multiple prescribing is not advisable in most cases. It can put the burden on consumers to decide, on their own, which of the multiple items they can afford to purchase. Prescribing inessential or ineffective medicines (such as cough syrups) is also a problem.

At the national level:

- A coordinated policy should be devised and introduced to promote the appropriate use of medicines. There should be an essential medicines list, developed according to established international practice and reflecting the health needs and priorities of a given country. The essential medicines list should also be in line with evidence-based standard treatment guidelines. The standard treatment guidelines should also provide the basis for practice, as well as for teaching and evaluating health professionals.
- Hospitals should set up medicines and therapeutics committees.
- The essential medicines list and standard treatment guidelines should be the basis for ongoing monitoring of the manner in which medicines are used. Appropriate and regular development and modification of the standards should reflect current knowledge and country-specific challenges and responses. Similarly, dispensers need to be trained according to these standards and their performance needs to be monitored.
- It is vital to provide reliable information on medicines and their use, both during the education of professionals and on an ongoing basis during their professional careers (such as through the publication of formularies, standard treatment guidelines, and regular prescribing bulletins). The information provided by manufacturers and importers may supplement this, but measures should be taken to ensure that it adheres, at a minimum, to *WHO Ethical Criteria for Medicinal Drug Promotion* (WHO 1988).
- The processes of prescribing and dispensing should, whenever possible, be separated to avoid overprescribing because of financial incentives to the prescriber.

Use in the household must be understood from economic, social, cultural, and gender perspectives

- Patients should always be given basic information about the medications that are prescribed for them (including name, dosage, clear use instructions, and possible side effects). This approach will require sensitivity to patient population characteristics, such as accommodating different dialects and meeting the needs of largely illiterate populations.

At the international level:

- Donors and global agencies engaged in the health field need to work together to promote the appropriate use of drugs. The *WHO Ethical Criteria for Medicinal Drug Promotion* (WHO 1988) should be updated and extended to deal with newer issues, including the trend toward direct-to-consumer advertising and the increasing use of the Internet to promote medicines.
- WHO should also ensure the worldwide sharing and dissemination of authoritative texts on the best means of treating major and epidemic conditions, so that these can form the basis for national guidelines.

Better use of medicines in the home. To ensure the well informed use of medicines in the home, long-term and incremental behavior change is needed. Education and culturally appropriate information on the use of medicines must be made available through appropriate channels, with special consideration for illiterate and minority-language populations. It is often the case that even medicines that have been appropriately prescribed and dispensed are still improperly used in the home. For example, the consumer may not have received adequate information about the treatment and the labeling could be inadequate. She or he may not be able to read instructions. Consumers may seek a savings, by stopping treatment when they feel better because they believe that the saved doses can then be available for use in future illness. A need exists to understand use in the household from economic, social, cultural, and gender perspectives. Information that is primarily technical in nature could be missing the point of why inappropriate use is taking place and consequently could be of little use in changing behavior.

At the national level:

- Governments should seek to educate the public on priority health issues, including the proper use of medicines. This general information should be supplemented by medicine-specific information, disseminated to households or patients in a culturally appropriate manner. This should not be a unilateral task for the authorities: community mobilization around issues of health and education is common. Forming alliances with community groups will be a valuable way of disseminating important information.
- As is the case with commercial promotion to professionals, pharmaceutical advertising directed to the general public should adhere to accepted standards and be responsive to local concerns.

Curricula for all workers who prescribe and dispense should be upgraded

At the international level:

- International health organizations and NGOs should continue to develop and disseminate health literacy information related to appropriate use of medicines for use in developing countries.

Crosscutting issues

The Working Group on Access to Essential Medicines identified two key crosscutting issues:

- The persistent and often worsening loss of skilled health workers is a threat to all efforts to improve health systems, including access to medicines.
- Gender is a key determinant in who gets access to medicines, why, and how. The extent to which gender considerations are integrated in policies and programs affects their success.

Human resources. Needed improvements in medicines supply, distribution, prescribing, and dispensing are not going to be realized if the entire underlying issue of human resource requirements is not adequately and urgently addressed. At its simplest, many more skilled workers need to be trained, deployed, and retained in the healthcare system. However, as studies of human resource issues in developing countries show, the problems are daunting and complex. New approaches and substantial resources will be required.

At the national level:

- Healthcare workers need to be paid wages that will ensure they can work in the field of their training.
- Governments should develop programs that will increase the sheer numbers of qualified workers and also ensure improved distribution, especially to poorer and rural areas. In many indebted developing countries, social sector spending limits continue to impede the country's ability to be responsive to health staffing needs.
- Curricula for all healthcare workers involved in prescribing and dispensing should be progressively upgraded and continuing education provided.
- The community's own resource persons should be mobilized to participate in healthcare planning and delivery of large-scale treatment programs (such as vaccines programs).

At the international level:

- Important support can be provided for training professionals, using internationally tested curricula.
- Donor financing should be available to subsidize staff wages in critical need areas.
- The brain drain of all types of health professionals from developing to developed countries is becoming a real crisis in some countries, such as Ghana, South Africa, and Zimbabwe. The international community needs to highlight the problem and reach consensus about how to

Medicines information should be gender responsive and made available in ways that are useful to women, who are often the primary care providers

reduce and manage the impact of this migration on developing countries. Possible solutions include banning active recruitment of health workers from developing countries or reimbursing training costs to the country that is losing that worker.

- International financing agencies, such as the World Bank and the GFATM, and major bilateral donors should focus on training and building capacity for a substantial number of supply chain managers and other essential health workers in developing countries.

Gender. Gender discrimination in all facets of women's and girls' lives has devastating consequences for their health and mortality. However, merely focusing on gender in isolation as a health issue will not succeed. The broader fundamental social, cultural, political, and economic interlocking roots of women's inequality in all societies must be tackled. Priority areas should include eliminating all forms of violence against women, especially sexual violence; improving economic security; removing discriminatory inheritance laws; and ensuring access to education for all girls.

At the national level:

- In health systems, policies and plans should mainstream gender considerations. This can be done only if women's participation increases and it is valued in decisionmaking.
- Governments should collect sex- and gender-disaggregated data on healthcare access and use, which, in combination with adequate gender analysis, should inform policies, plans, and budgets.
- National essential medicines lists should contain the core medicines and devices for sexual and reproductive health recommended by the United Nations Population Fund and WHO (see box 1.1).
- Access to healthcare and treatment must be significantly increased for women and girls if the Millennium Development Goals are to be on track.
- Policymakers and planners would benefit from having more research done on the gendered aspects of medicines access and use by women and girls and men and boys. Medicines information should be gender responsive and made available in ways that are useful to women, who are most often the primary care providers in families.

At the international level:

- UN agencies and the GFATM should adopt policies and approaches that ensure that gender considerations are adequately integrated into all aspects of their planning, activities, and budgets.

Conclusion

Access to medicines has always been an important concern in health development policymaking and programming. But it was WHO's call for "Health for All by 2000" in the 1977 Alma Ata Declaration that launched what has been

A key finding is the need to involve the community in developing health system policies and programming

an ongoing effort to examine and eliminate barriers to access, especially for the poor. Both the frameworks and the expertise exist to understand, in complexity and scale, how to address all of the major obstacles. However, to date, the world remains a long way from attaining equitable access within developing countries, let alone across regions.

Thirty years ago, medicines policy was a technical discourse mainly among UN agencies, ministries of health, and international experts. However, the growing AIDS pandemic has galvanized discussions about access to treatment. The United Nations, donors, recipient governments, and suppliers are being pressured by a growing global network of public interest NGOs and civil society groups that need medicines and are not able to get them. New bodies, such as the GFATM, have been founded to provide financing for national programs to tackle three of the major diseases of poverty. Existing organizations, both public and private, have become increasingly engaged in finding new ways to increase access to medicines. But more needs to be done, and it will require new thinking and new approaches.

In the last decade, most developing countries have undertaken measures to improve access to medicines, with varying degrees of success. Even where there have been setbacks, the experience gained strongly indicates that progress is possible. Where both the initiatives and the results have been monitored, lessons emerge that can be adapted to local conditions and applied elsewhere. A key finding is the need to involve the community in developing health system policies and programming.

Not all trends are developing satisfactorily. Finance is still seriously insufficient, and the overall health sector situation in developing countries remains extremely complex. In addition, the continued advance of AIDS in all of the poorest countries threatens to overwhelm already weakened, limited, and inequitable health systems.

According to WHO, access to essential medicines worldwide increased from roughly 2.4 billion to 4.3 billion between 1975 and 1999 (WHO 2004a). A closer look at the numbers, however, shows that the overall number of those without access remains relatively unchanged and that these people are primarily the poorest and most marginalized. Consequently, it remains to be seen if current knowledge about access and current approaches to increasing access adequately reflect a truly pro-poor framework from the global level (especially international financial institutions, UN agencies, and development strategies of major donor countries) to the local level (including real commitment by national governments to tackle poverty and take steps to improve national economic development).

Chapter 1

The problem

A very large part of the world's population has inadequate or no access to essential and life-saving medicines. Along with skilled and dedicated healthcare providers, medicines[1] are the most significant means that society possesses to prevent, alleviate, and cure disease. Most illnesses, especially infectious diseases, are either preventable or to some extent treatable with a relatively small number of medicines. Combined with appropriate public health interventions, appropriately prescribed essential[2] medicines and vaccines could, in principle, massively reduce the impact of disease on communities. Despite this fact, a large proportion of the world's population today still has either only limited access to appropriate medicinal treatment, or no access at all.

The consequences of this inadequacy include an enormous loss of life from preventable or treatable diseases (such as tuberculosis, pneumonia, acute respiratory infections, malaria, diabetes, and hypertension) and significant human suffering, particularly among the poor and marginalized populations of the world. The lack of access to life-saving and health-supporting medicines for more than 2 billion poor people stands as a direct contradiction to the fundamental principle of health as a human right. Illness is a major reason that the nearly poor slide into profound poverty. Illness decreases people's ability to work (be it remunerative or not). Illness orphans children and prevents them from getting the education they need. Women and children make up the majority of the poor, and their low status in many societies often means that they have even less access to medicines. Improving access to medicines must be a key component of strategies to strengthen healthcare.

Increasing access to medicines in developing countries, especially for the poor, offers many challenges. These can be crystallized into two main areas:

- How to increase access to affordable existing medicines in resource-poor settings, which countries can do by improving the selection and use of

The World Health Organization estimates that a third of the world's population lacks access to the most basic essential medicines

essential medicines, taking steps to ensure affordable prices, increasing sustainable financing, and strengthening reliable supply systems.
- How to find new ways to promote the development of new medicines and vaccines to treat diseases of poverty.

The World Health Organization (WHO) estimates that a third of the world's population lacks access to the most basic essential medicines, while in the poorest parts of Africa and Asia this figure climbs to one-half (WHO 2000f, WHO 2000a). WHO has also estimated that, in Africa and Southeast Asia, prompt diagnosis and treatment with appropriate medicines could save approximately 4 million lives annually (DFID 2004b). Moreover, it is often the poorest who are paying the highest out-of-pocket expenses for medicines because the public sector in developing countries is unable to provide affordable medicines reliably. Medical insurance schemes cover less than 8 percent of the population in Africa (WHO 2004a), and these schemes may not cover prescription medicines on an outpatient basis. Consumers often judge health systems primarily by whether or not they get medicines when they seek treatment. Participatory assessments during national poverty reduction strategy processes often find that the availability of medicines is a primary indicator of the effectiveness of healthcare delivery.

Though access to essential medicines has improved in recent years, WHO reports that delivering "the right medicines to the people who need them at the time they need them remains a major challenge" (WHO 2004a). The analysis contained in the WHO's 1999 *World Medicines Situation* showed that roughly two-thirds of the world's population have regular access to essential medicines, up significantly from 1975, when this proportion was just under one-half (WHO 1999). However, global population growth has meant that the absolute number of people without access has remained nearly constant, at approximately 1.7 billion.

The lack of access to medicines in most developing countries reflects both the lack of sufficient incentives for the development of new medicines to target those communicable diseases that disproportionately afflict the poorest countries, as well as the inability to pay for and effectively distribute those that do exist. The result is what the U.K. government has called a "mismatch between pharmaceutical needs in developing countries and the current nature of the global pharmaceutical market" (DFID 2004a, p. 14).

Tuberculosis (TB), once regarded as a receding public health problem that had become containable in the developed world, provides a clear example of the challenge of reemerging diseases (WHO 2004b). It is a well documented threat:

> [T]oday TB is making a comeback. One in three people in the world are infected with the Tuberculosis *bacillus*—they have latent TB. Normally only a small proportion—roughly eight million people per year—of these progress to the clinical disease known as active TB, in the vast majority of cases, characterized by a lung infection. Those with active

The world has the aggregate financial means to provide all people with basic medicinal treatment, but those means are not evenly distributed

> pulmonary TB are the most likely to spread the TB bacilli to others. . . . TB kills roughly two million people every year. Around 95 percent of all patients with active TB live in the developing world, where 99 percent of all TB deaths occur. Exact data are hard to come by, but at least 4 percent of all TB patients worldwide are resistant to at least one of the current first-line drugs. In parts of Eastern Europe, nearly half of all TB cases resist at least one first-line drug. Multidrug-resistant (MDR) TB, defined as resistance to at least rifampicin and isoniazid, the two most powerful TB drugs, might be spreading as fast as by 250,000–400,000 new cases each year (Dye and others 2002).[3] Their treatment relies on second-line TB drugs that have far lower efficacy and require even longer administration periods (18–24 months)—with much higher cost and much higher rates of adverse effects (MSF 2004a, p. 3).

The response to the global TB pandemic requires innovative technology: new medicines, new diagnostics, and new vaccines. It also requires innovation in the ways that new medicines are developed and made available in the regions bearing the heaviest TB burden. The Working Group on TB (one of the other working groups within the UN Millennium Project Task Force on HIV/AIDS, Malaria, TB, and Access to Essential Medicines), has identified the issue of new medicine development as a critical focus of analysis and action. The Working Group on TB recommends the following steps to increase access to TB medicines in their final report (UN Millennium Project 2005b):

- Donor agencies should increase their investments in the research and development (R&D) of new and affordable TB diagnostics, medicines, and vaccines through public-private partnerships, and in related projects such as the strengthening of clinical trials capacity in DOTS programs.
- The Stop TB Partnership and WHO should advocate for these investments to complement their access and treatment efforts. Advocacy should be based on a clear definition of the economic and social justifications of investing in new tools.
- Regulatory agencies should assist in harmonizing streamlined regulatory requirements to introduce new TB diagnostics, medicines, and vaccines.
- The network of Stop TB partners should work to ensure that new tools respond to the greatest demands of users, and should expedite testing and rollout in high-burden settings.

The most frequently cited cause of inadequate access to medicines is that individuals cannot afford them. The world has the aggregate financial means to provide all people with basic medicinal treatment, but those means are not evenly distributed (Attaran and Sachs 2001).[4] Even an adequate national financial base is no assurance that these health issues will be adequately prioritized within countries. For example, a country with very limited resources may prioritize primary education over health services; a country may make military spending

In 1999, roughly 80 percent of the global population without access to essential medicines was living in low-income countries

a priority over social sector financing; a country may fail to tackle corruption. Most developing countries demonstrate substantial internal inequalities; these can be exacerbated as they strive to fulfill repayment obligations and meet conditionalities of powerful international financial institutions, such as the International Monetary Fund and the World Bank. Finally, within the medicines sector itself, available resources may be poorly managed or misappropriated. The fact that medicines are items with a small volume, high unit value, and universal demand makes the sector susceptible to pilferage and corruption.

Definition of access to essential medicines

The Working Group on Access to Essential Medicines incorporated the WHO definition that provided the basis for the Millennium Development Goals indicator used to assess access: the proportion of population with access to affordable, essential drugs on a sustainable basis is the percentage of the population that has access to a minimum of 20 of the most essential drugs. Access is defined as having drugs continuously available and affordable at public or private health facilities or drug outlets that are within one hour's walk of the population (WHO 1999b). The working group qualified accessibility to apply only to products that are effective and of consistently good quality, that have no financial obstacle to a patient receiving it, and that have available the knowledge and guidance needed to use them properly.

Despite the progress made in the last decades, the likelihood of a person having access to essential medicines is still affected greatly by that person's income level. The *World Medicines Situation* found that people in poorer countries were much less likely to have access to these medicines (WHO 2004a). According to WHO, in 1999, roughly 80 percent of the global population without access to essential medicines was living in low-income countries (figure 1.1). This is a disproportionate share of the global burden, given their estimates that low-income countries account for approximately 60 percent of the world's population (WHO 2004a). In contrast, only 0.3 percent of those lacking access to essential medicines lived in high-income countries, which account collectively for about 15 percent of the world's population. In a global context,

Figure 1.1
Distribution by country income group of people without access to essential medicines, 1999

Source: WHO 2004a.

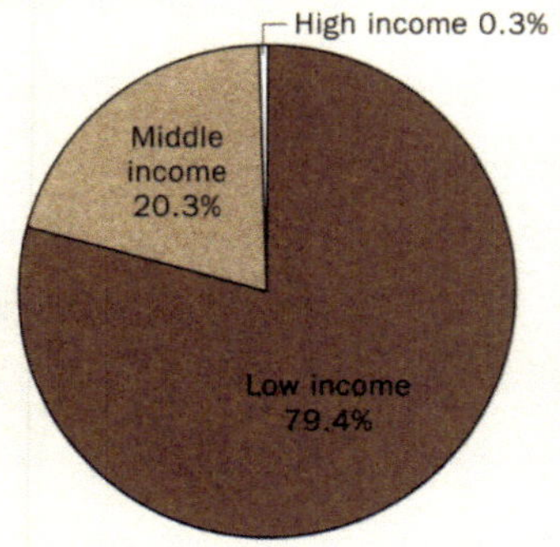

Together, India and Africa account for 54 percent of the world's population without access to essential medicines

that 15 percent of the world's population consumes 91 percent of the medicines produced (WHO 2000a). Of people living in low-income countries, nearly 40 percent did not have access to essential medicines in 1999 (WHO, 2004a).

Geographically, the lack of access to essential medicines is especially severe and concentrated in Africa and India (figure 1.2). In fact, 38 percent of the people without access to essential medicines live in India. Another 15 percent of the people without access live in African countries (WHO 2004a). Together, India and Africa account for 53 percent of the world's population without access to essential medicines (WHO 2004a). Although the disease burden and mortality from preventable or curable illness is highest in African countries, pervasive poverty means that the continent's share of the global pharmaceutical market is only slightly more than 1 percent.

India's and Africa's inordinate share of the global population without access is not entirely a function of population. India accounts for only 17 percent of the world's population. Similarly, Africa has roughly 10 percent of the world's population. This translates to very high absolute numbers of people without access in these two regions. Sixty-five percent of Indians and 47 percent of Africans lack access to essential medicines (figure 1.3), while the equivalent proportion is 14 and 22 percent in Europe and the Americas, respectively (WHO 2004a).

The lack of access to medicines throughout large proportions of the populations of most developing countries reflects both the lack of sufficient incentives for developing new medicines to target the communicable diseases that disproportionately afflict the poorest countries, as well as the inability to pay for and effectively distribute those that do exist. The result is what the U.K. government has called a "mismatch between pharmaceutical needs in developing countries and the current nature of the global pharmaceutical market" (DFID 2004, p. 14).

When examining access to essential medicines for the poor, the Working Group on Access to Essential Medicines identified a number of fundamental problems common to many countries. What is very clear, however, is that the basic knowledge and technical information already exist to increase access to

Figure 1.2
Distribution by region of people without access to essential medicines, 1999

Source: WHO 2004a.

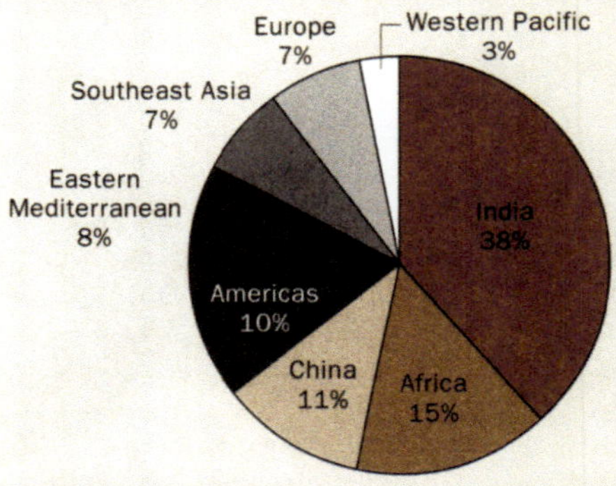

Figure 1.3

Share of people without access to essential medicines by region, 1999

Source: WHO 2004a.

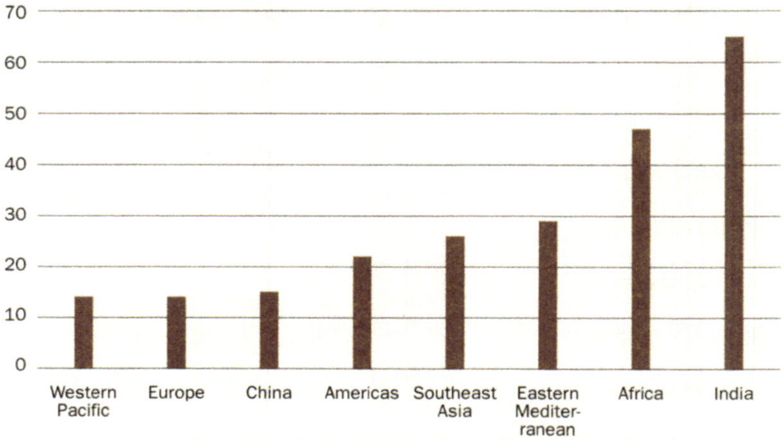

all segments of a population. Furthermore, the world possesses the resources to fund adequate access to essential medicines and functioning health systems in the developing world.

Ideally the working group would have wished to propose a simple approach to improving access to medicines. Because obstacles to access are many and diverse, and because they differ in nature and degree from one country to another, this is not possible. The issues surrounding access are complex, at times culturally specific, and often fluid; the solutions can be no less. Some will need to be applied at the global level, while others need to be selectively employed in particular countries or regions depending on the situation and its context.

Overarching barriers

At the risk of overshadowing the other vital issues presented in this report, the Working Group on Access to Essential Medicines would however point to six of the most important barriers to access that merit special attention and action at this time. The first four relate to barriers to existing medicines, while the last two constitute barriers to the development of affordable and available new medicines and vaccines.

Barriers to existing medicines

1. *Inadequate national commitment* to making healthcare a priority from the national to the local levels remains one of the greatest barriers to increasing access to existing medicines. There are many reasons for this lack of prioritization. Key among them are a lack of political will by policymakers to make the needs of the poor a priority; donor programs that can skew or limit national governments' abilities to set health policy; debt servicing and conditionality for loans from international financial institutions that can further limit government responsiveness to basic social service needs of citizens; and, unfortunately, the threat of corruption that continues in the healthcare sector at all levels.

Most poor countries will require significant donor funding to achieve universal access to essential medicines

2. *Inadequate human resources* for health, including pharmacists and pharmacy technicians, is a growing problem that, if unaddressed, threatens to undermine all efforts to strengthen health systems and improve healthcare in much of the developing world. Education, information, and in-service training remain potent tools to change that situation. More needs to be done to identify what is needed to retain skilled workers, especially in the face of mounting demands for health workers, such as nurses and pharmacists, in developed countries. Retention plans and compensation schemes for countries that lose health workers should be investigated.
3. The *international community has not provided adequate finance nor consistently fulfilled its existing promises to developing countries.* Some proposed actions have not been carried out at all and others have not been carried out effectively. To achieve progress, there will be a need for political will, in both industrialized and developing countries, as well as a need for transparency on all fronts. Above all, there will be a need for increased levels of long-term financial support from the world community. It remains an unfortunate ongoing reality that some of the world's wealthiest countries remain the farthest from achieving their longstanding commitment to the international development assistance target of 0.7 percent of gross domestic product (GDP).
4. A persistent *lack of coordination of international aid* reduces access to medicines. Most poor countries will require significant donor funding to achieve universal access to essential medicines. They will also need much better aid coordination to avoid unnecessarily heavy reporting requirements and to avoid resource-wasting duplication of efforts. Sectorwide approaches should be used to promote improved coordination. Donors should commit aid that strengthens existing systems, that proactively targets the poorest and rural areas, and that avoids vertical programming by disease or by a given donor. A need exists at both the international and national levels for a great deal more transparency and coordination of effort between the large number of organizations that have already become involved in one way or another in this field. The involvement of so many bodies can and does lead to duplication of effort and to waste, and both are unacceptable. In some situations, there is every reason to merge complementary ventures. Pharmaceutical companies can and should contribute in their own particular way to the advancement of national medicines policies and the development of capacity in this field.

Barriers to the development of affordable new medicines

5. *The Trade-Related Aspects of Intellectual Property Rights (TRIPS) agreement may block access to affordable new medicines and vaccines.* After January 2005, generic production in India, the source of many vital existing medicines for developing countries without productive capabil-

WHO should be mandated to monitor the impact of TRIPS compliance by major developing country exporters on access to medicines

ities, will be fully subject to TRIPS provisions (WTO 1994). Concerns also exist that the August 30, 2003, decision reached by the WTO General Council concerning a waiver for TRIPS Article 31(f) (which would allow a compulsory license to be issued by the country in need and by the country that can produce the medicine for export) will be too cumbersome for developing countries to exploit (WTO 2003). Finally, the growing number of bilateral and regional trade agreements with major trading partners, such as the United States and the European Union, may often contain provisions that limit developing countries' use of existing flexibilities under TRIPS to protect public health (such as restrictive compulsory licensing conditions and parallel importation provisions, extended data protection, and forcing medicines regulatory agencies to take on national patent office oversight duties).

A major recommendation of this working group is for the WHO to be mandated, perhaps in coordination with the WTO or other trade bodies, to monitor the impact of TRIPS compliance by major developing country exporters and, in particular, to monitor the use of the August 30 decision over the coming two years as it concerns access to medicines. A report to the World Health Assembly and the WTO General Council, with recommendations, should be delivered no later than the end of 2007.

6. *The current incentive structure is inadequate to promote research and development of medicines and vaccines to address priority health problems of developing countries.* For a number of the most neglected diseases (such as African trypanosomiasis, Chagas disease, leishmaniasis, and dengue fever), which occur primarily in developing countries, new medicines need to be developed (WHO and IFPMA 2001). For others, new medicines are needed to address shortcomings of existing treatments, such as safety, efficacy, appropriate dosing, length of treatment, and the ongoing threat of drug resistance. Despite progress in funding research and development (R&D) for new medicines for neglected diseases, with notable contributions from philanthropic foundations and some governments and pharmaceutical companies, more financial resources need to be mobilized in a sustainable way to create a strong and sustainable pipeline of new products. New thinking, different means of financing and organizing medicines development, and other reforms are needed. For example, the WHO Commission on Intellectual Property Rights, Innovation, and Public Health should examine alternative international models to the current patent-based system for priority setting and financing of health R&D.

Finding solutions

Since the issues surrounding barriers to access have many causes, a single solution to improving the provision of medicines cannot be expected to succeed;

National medicines policy cannot succeed in isolation from broader health policies and government policies

it must be complemented by others. What this means is that every developing country should have an overall national medicines policy and strategy founded on the essential medicines concept. The aim of such a policy has been succinctly defined as existing to ensure that "safe and effective drugs of good quality are available and affordable to the entire population and that they are rationally used" (World Bank 1993). The most important step is to develop a model that meets national needs and to employ it as a basis for developing and managing the system as a whole.

National medicines policy cannot succeed in isolation from broader health policies and government policies in general. A ministry of health is unlikely to succeed in this area unless it has clear and acceptable understandings reached with other government departments dealing with such matters as finance; the training curricula for health professionals; the salaries of public employees; and practices regarding trade, taxation, and customs duties, all of which are likely to have positive or negative impacts on the supply of medicines.

Health sector strengthening and development to reach the UN Millennium Development Goals should be done in the context of the national poverty reduction strategic planning being adopted in poor, indebted countries. A recent WHO review of national Poverty Reduction Strategy Papers (PRSPs) and health pointed out that much good information is being generated about health system needs, including access to medicines, in analytical phases of the PRSP process (WHO 2004a). However, it would appear that when government responses were formulated, there was a tendency to rely on existing health policy approaches and budgets, which rarely included any community or civil society participatory processes. Instead, they tended to reflect top-down prioritization shaped by international financial institution conditions and a lack of political commitment to reorient government focus more toward the social sector. A need exists, therefore, to examine how health sector and other social sector needs can be better addressed in the poverty reduction strategy planning processes, including how information and needs assessments gained in participatory analysis can be better translated into government planning and budgets.

The goal of achieving universal access to essential medicines in developing countries will require creating and strengthening primary health systems along with the myriad specialized administrative and functional features needed to maintain a sustainable supply of all essential medicines. Firm priorities must be set, including an urgent need to reassess the importance of health sector investments by donors and recipient governments and to increase resources substantially through greater political will on both sides.

The lack of adequate and accessible pharmaceutical treatments for HIV/AIDS, malaria, and TB has received wide attention. But the problem extends to many other illnesses, and it is much wider in scope because it emerges from the failure to provide adequate healthcare overall. Poverty,

Access to medicines merits urgent study because it is a challenge that can be solved

lack of political will, social disruption, inconsistent aid and trade policies, and weak institutional supports have left poor countries with fewer tools to fight chronic, acute, and newly emerging and remerging illnesses. Overall, attention and innovation has been concentrated on diseases found predominantly in industrialized countries.

This situation—which is worst among the urban poor and the populations of isolated rural areas—has serious repercussions for human health and welfare. It will also have a profoundly negative impact on local, national, and regional economic growth and the general development of societies (CMH 2001). Widespread displacement of large populations, internal and international migration, and global exchange and commerce are features of globalization in the twenty-first century. The failure to address infectious diseases such as multidrug-resistant (MDR) TB systematically will ensure that their patterns of transmission will evolve from relatively circumscribed areas to global dispersion.

Context

Access to medicines merits urgent study because it is a challenge that can be solved (Quick 2003). Its various causes are known, and many of them can be eliminated with adjustments in government policies; economic, scientific and commercial priorities; and individual behavior with respect to medicines. These are attainable objectives, if sufficient focus is paid to them. UN Secretary-General Kofi Annan (Annan 2001) has defined a specific strategy of engagement through the framework for implementing the UN General Assembly's Millennium Declaration; this approach and others are considered in this report. It should be noted that access to essential medicines is closely linked to the achievement of other Goals and should be seen in that broader context. Table 1.1 outlines some of the main linkages between access and some of the other key Goal strategies elaborated by the UN Millennium Project.

Human right to health

The promotion and protection of human rights are principal mandates of the United Nations. The earliest global statement on human rights and health is in the WHO constitution, adopted in 1946. This includes the following relevant passages:

> The States parties to this Constitution declare, in conformity with the Charter of the United Nations, that the following principles are basic to the happiness, harmonious relations and security of all peoples:
>
> - Health is a state of complete physical, mental and social well-being and not merely the absence of disease or infirmity.
> - The enjoyment of the highest attainable standard of health is one of the fundamental rights of every human being without distinction of race, religion, political belief, economic or social condition.

Table 1.1
Key intersections between access to medicines and other Millennium Development Goal themes

Health systems	Consistent supplies of safe, reliable medicines are critical to the success of health interventions. Patient confidence and trust in health services are frequently and heavily influenced by the reliable availability of essential medicines at the point of care. The creation of new (and expansion of existing) primary services is required, including dispensing free or extremely low-cost medicines, with particular focus on poor people and geographically isolated communities.
Trade	Monitoring international, regional, and national trade, tariff, and intellectual property policies is required to ensure that the production, importation, and use of essential medicines is not obstructed.
Education	Education and literacy programs promote individual and collective empowerment and increase awareness of ways to prevent and treat a wide range of diseases, acute illnesses, and preventable injuries. These programs should include the appropriate use of over-the-counter and prescribed medicines. Expanded access to higher education will increase the number of healthcare workers (including pharmacists) at all levels of care.
Gender equality	Women's empowerment leads to more informed and focused demand for available treatment options. Women are often the first line of care for their families. Outreach on the safe use of medicines that is targeted to women can benefit large numbers of family and extended family members.
Slum upgrading and urban planning	Slum upgrading and urban planning improves access to general health services and disease-specific interventions.
Science and technology	Research should increasingly target the discovery and development of new and effective medicines, vaccines, and diagnostics to prevent and treat diseases associated with poverty. Improved communication technology can facilitate efficient procurement of medicines and inventory control of stock.
Energy	Reliable, sustainable, and modern energy services simplify and greatly improve the quality of diagnostic and treatment services across the healthcare spectrum.
Transport	Improved transport infrastructure reduces the cost of medicines and diagnostics and critical supply distribution. Improved geographic access, especially in isolated rural areas, will facilitate access to care. Improved transport infrastructure can facilitate movement of healthcare workers and reduce systemic obstacles to working in rural areas.

The right to health was subsequently recognized in a series of other global instruments. The Universal Declaration of Human Rights affirms that "everyone has a right to a standard of living adequate for the health of himself and his family, including . . . medical care" (United Nations 1948). The International Covenant on Economic, Social, and Cultural Rights, adopted in 1966 and ratified by 147 states, and other international and regional treaties recognize "the right of everyone to the highest attainable standard of physical and mental health" (UNHCHR 1966). In May 2000, the Committee on Economic, Social, and Cultural Rights specifically elaborated that Article 12

By 2001, all 193 countries in the world had signed at least one international or regional treaty that confirms the right to health

of the covenant included the right to essential medicines (WHO 2002a; Hogerzeil 2003). In 2002, the International Guidelines on HIV/AIDS and Human Rights (OHCHR and UNAIDS 2002) were updated to address the right to medical treatment, including access to medicines, as a specific right.

The right to healthcare facilities, goods, and services mentioned in the covenant includes a right to appropriate treatment of prevalent diseases, preferably at the community level. The covenant provides for progressive realization of the right to health and acknowledges the limits of available resources. It states that parties have an immediate obligation to guarantee that the right to health will be exercised without discrimination of any kind, and to take deliberate and concrete steps toward its full realization, with emphasis on vulnerable and marginalized groups such as children, women, and elderly people. By 2001, all 193 countries in the world had signed at least one international or regional treaty that affirms the right to health; more than 90 countries have included it in their national constitutions.

Increasingly, there is a consensus that human rights should incorporate the ability of individuals to maintain and restore good health through access to at least a basic level of primary care, including essential medicines. WHO, the World Bank, and major international development NGOs all promote rights-based approaches to poverty reduction and health development. However, the right to health is not reflected in the current global situation, in which entire populations—particularly the poor and underprivileged—commonly have little or no access to essential medicines or other basic health services.

Role of gender and women's status

At various points in this report it will be necessary to consider the extent to which gender issues affect drug access.[6] Understanding why more women and girls are infected with and affected by HIV and AIDS, comprehending the fact that many women do not and cannot seek treatment or buy medicines for their illnesses, and even explaining the manner in which medicines ordinarily enter the home, demands a consideration of the role of gender and the need for change in its construction. No health intervention can be completely successful without gender analysis of the problem and—crucially important—adequate translation of that analysis into strategies, financing, and implementation.

Gender analysis involves a systematic examination of the socially constructed roles played by women and girls and men and boys, and the power relationships arising from and reinforced by them. It includes looking at the division of labor, productive and reproductive activities, and access to and control over resources and benefits, and examining the socioeconomic and environmental factors that influence women and girls and men and boys differently. Such an analysis requires gender-disaggregated data (such as sex as an independent variable and information on how labor is divided and valued). Gender analysis

In India and China, 75 percent of those without access to essential medicines are women and children

takes into account the variable and intertwined factors of race, class, ethnicity, sexual preferences, and age.

Gender analysis also takes into account a woman's role in the household. "According to the World Bank, women, although they are not recognized as health care workers, are in fact responsible for between 70 and 80 percent of all health care provided in developing countries" (Bisilliat 2001, p. 17). This role extends to the use of medicines in the home. Yet within a household, women responsible for family healthcare and decisionmaking about care often do not control household money, so they may not be able to access needed finances to get transport to a facility or pay for treatment. In some societies, they may not be allowed to seek treatment unescorted by a male family member, which is a constraint on treatment-seeking behavior. One challenge that women face is that their healthcare needs differ from those of men and boys because of their reproductive roles. One aspect of gender inequality is that women may be limited to accessing only those healthcare services directly related to maternity or childcare. This neglects a broad spectrum of life-span health issues that will also arise.

The inequality between women and men has never been more starkly evidenced than by the link between violence, economic deprivation, social roles, age, and the risk to women and girls of being infected with HIV. The Joint United Nations Programme on HIV/AIDS (UNAIDS) estimates that 58 percent of adults infected with HIV in Sub-Saharan Africa are women (UNAIDS 2002). UNAIDS has highlighted the urgent need to tackle women's inequality and to have gender-sensitive health responses if progress is going to be made against the pandemic. A gender-blind HIV and AIDS intervention is assuredly a key indicator for failure. AIDS has stretched traditional caregiving roles to the very limit. Notably, HIV-positive women retain this responsibility, even when they themselves are sick.

Concerted efforts must include increasing access to treatment for girls and women. According to WHO, of the 50 percent of the populations that do not have access to essential medicines in Africa and Asia, 75 percent are women and children (Bisilliat 2001, p. 17). According to a survey of antiretroviral medicines use by Oxfam in Uganda (Oxfam 2002), "Women are severely disadvantaged in gaining access to this life-saving treatment." The survey showed what has been well established concerning other health decisions: if choices have to be made, women are denied access in deference to others in the family. Poverty and this household discrimination explain why many women are not tested, even when the partner is HIV-positive. It is accepted that they would not be treated, even if their status were known. The Oxfam report provides a useful example of the untenable choices people are forced to make every day:

> John, a local fisherman near Kampala, has felt much better since he started antiretroviral therapy. He can work and provide for his family. John's wife has also recently been diagnosed with HIV. "I do not have the money to

pay for medicines for both of us." If he pays for his wife, his children cannot go to school and "they will not have a future" (Oxfam 2002, p. 2).

Recent WHO country-level surveys of the pharmaceutical sector show that inadequate labeling is a problem, especially in rural areas, in a number of countries. Illiteracy rates in developing countries normally are higher for women, due in large part to the discrimination against girls in accessing education, especially in poor households. So even with proper labeling, the inability to read the instructions can create a health risk. Moreover, illiteracy means that broad campaigns using printed medical educational materials will be of no practical use to the most vulnerable populations.

Medicines policies and programs at country level need to be gender responsive. This means that they must be based on an adequate gender analysis that informs strategies, plans, and budgets. It means that women must be consulted and substantively involved in health sector planning. For example, the revision of essential medicines lists should include, at a minimum, the recommendations from the United Nations Population Fund (UNFPA) and WHO for contraceptives (box 1.1).

Box 1.1

Essential reproductive health medicines and supplies

Source: Adapted from UNFPA and WHO 2003.

UNFPA and WHO have collaborated on a list of drugs and commodities to be procured for key needs in the area of reproductive health.

The list addresses various needs to provide:

- A full range of contraceptives (such as hormonal methods, intrauterine devices, barrier methods, and subdermal implants).
- Maternal and neonatal healthcare commodities (anesthetics, analgesics, antibacterials, anticonvulsants, antiseptics/disinfectants, oxytocics, vaccines, and vitamins, among others, including equipment for emergency obstetric care).
- Reproductive tract infection commodities (diagnostic tests and first-line drugs).
- HIV prevention commodities (male and female condoms and drugs to prevent mother-to-child transmission).
- General equipment and supplies for providing primary and secondary healthcare (sterilizing equipment, lighting, anesthetics and their equipment and supplies, oxygen, postoperative medication, operating theater equipment, ward equipment, drugs for gastric acidosis, muscle relaxation, and anaphylaxis, and cholinesterase inhibitors).

The list of essential drugs and other commodities for reproductive health services concentrates on those drugs that are necessary at the primary healthcare level. The appendixes provide detailed specifications. However, for certain components of reproductive healthcare, most notably maternal care, it is absolutely crucial that services are accessible at the first referral level and the necessary commodities for this level have also been included where appropriate. The list is not intended to be comprehensive; however, it includes those items that are most important for providing the core components of reproductive health primary care.

The list is principally intended for people involved in planning and implementing reproductive health programs, including program managers and development assistance agencies and organizations.

In purely economic terms, the return on investments in essential medicines is substantial

Essential medicines lists should not be gender blind. For example, the WHO model essential medicines list includes only condoms, without mention of female versus male condoms or sizes. Nor should essential medicines lists or procurement policies exclude treatments for women based on political, cultural, or religious perspectives of policymakers (such as contraceptives or medical abortifacients). National medicines policies need to recognize and reflect gender analysis in policy formulations. Monitoring and evaluating health policies(including medicines policies) should have both sex- and gender-disaggregated indicators.

To help meet the Millennium Development Goals, all parties engaged in developing new medicines needed to fight diseases of poverty should ensure that these processes and the medicines that result from them are gender responsive. For example, contraceptive implants (provided they are safe), can protect women who might be beaten by partners who knew that they were using family planning. A fixed-dose combination antiretroviral increases ease of compliance, thereby reducing fears of one's status being exposed in the family or community. More needs to be known about the differential responses to medicines based on biological differences between women and men. Research methodologies and clinical trials need to be gender responsive.

Economic dimensions

The investments needed to ensure access to medicines for all are often portrayed as an insurmountable hurdle. In fact, since essential medicines alleviate the burden of disease and reduce morbidity and mortality, universal access to them would yield significant and measurable economic returns through improved health outcomes. When assessing the merits of a particular project to improve health, the costs of resolving access barriers can often be set directly against the anticipated or proven returns in terms of the amount saved on healthcare and disability. In purely economic terms, the return on investment is substantial (Rankin 2003). Viewed in this way, essential medicines more than pay for themselves. The World Bank's pioneering report *Investing in Health* delineated the economic effects of improving the health of populations, and some of the most striking examples it provided related to medicines and vaccines (World Bank 1993).[7] The loss of disability-adjusted life years in Sub-Saharan Africa in 1990 alone totaled 7.4 million years for TB and 16.1 million years for malaria, both diseases, in principle, being curable by medication. Treatable childhood diseases, such as polio, measles, and pertussis, account for 5.2 percent of disability-adjusted life years in high-mortality, low-income countries (WHO 2002g), even though vaccines for these diseases have existed for at least 50 years.

A study of TB in Eastern African countries showed that the cost per life-year saved was between $1 and $3. Improving people's health and prolonging lives translate directly into gains in worker productivity, improved use of natural resources, reduced costs of medical care, and other economic benefits,

The amount that a country spends on medicines ought to be decided in a manner consistent with clearly articulated health policy priorities

some of which are directly measurable (Brazil, Ministry of Health 2001).[8] Out-of-pocket costs of medical treatment, including medicines, combined with loss of work due to illness, frequently reduce disposable incomes and deepen levels of poverty (CMH 2001).[9]

In the poorest developing countries, the money to fund health services and to provide medicines is simply not available.[10] Moreover, in these countries, donor support underpins substantial portions of national budgets. In Uganda, for example, donor financing accounts for 52 percent (see also appendix 1). For these countries, the inescapable fact is that increased donor support will be required to provide even a minimally sufficient primary care package.

However, some developing countries have aggregate national resources sufficient to meet all the primary healthcare needs of their citizens, yet non-health priorities are given precedence. The WHO Commission on Macroeconomics and Health (CMH) demonstrated that many middle-income countries already have the internal resources needed to raise the level of financing for a minimum universal package of health interventions, including access to the most essential medicines. For others, including many in Sub-Saharan Africa, significant additional donor funding will be required to provide a minimum package (CMH 2001).

The amount that a country spends on medicines ought to be decided in a manner consistent with clearly articulated health policy priorities. Actual per capita outlays on medicines can—and should—differ from one country to the next. For the poorest, most heavily indebted countries, an expedited, expanded system of debt relief will prevent the added value of additional donor support from disappearing into the quagmire of effort to meet impossible loan repayments. Once a country has advanced on the road to development, the key issue will be one of efficient allocation, balanced by the exercise of political will and the empowerment of ministries of health.

Cultural dimensions

There are culturally based differences between and within countries that will determine access to medicines and the utilization of healthcare in general. In addition to gender issues, other culturally specific factors include class-mediated relationships; patient and physician compliance; traditions of authority, dominance, and resistance; variable degrees of illness stigmatization; ethnicity; generational shifts; and culturally mediated perceptions of illness, health, and healthcare interventions. The history of a community also plays a major role in determining how access might best be supported.

Increasing the availability of essential medicines in settings in which biomedical treatments have been either absent or only nominally available carries the responsibility of providing appropriate training and educational resources to both healthcare providers (including pharmacists and pharmacy technicians) and the public. The commercial advertising and promotion of medicines differs both in its approach and in its effects from one culture to another. It calls for regulatory vigilance in regions that have not previously been exposed to these commercial pressures and have therefore not had an opportunity to develop a critical approach to them. Regulations, prescriber practices, dispensing systems, and patterns of home use must also be part of an evaluation of prescribing and consumption practices.

The case study in appendix 1 on access to medicines in Uganda provides a snapshot of this issue's intersections with poverty. By providing a context to the issue of access to medicines within the larger socioeconomic frame in which it is occurring, the case study seeks to demonstrate the interrelated nature of the obstacles and challenges facing Uganda today and to propose some solutions targeted to addressing access to medicines.

The role traditional medicines play in complementing Western medicine is often overlooked in discussions of access

Traditional medicines

The role traditional medicines play—and could play in the future—in complementing Western medicine and relieving the burden on the health services has often been overlooked in discussions of access. Many developing countries have a rich flora and a long tradition of using these plants to prepare medicinal products. In these same settings, a large proportion of the population relies heavily on traditional practitioners using their knowledge to make and provide natural medicines. Traditional healers commonly enjoy a greater degree of popular trust than the official health services and often provide care to those unable to provide monetary payment; payment in kind is often the standard. There seems little doubt that many of the remedies used in traditional medicine have at least symptomatic and sometimes curative value. Some, ranging from digitalis to vinca, have in the last two centuries been assimilated into Western biomedicine for this reason. The fact that traditional healers are trusted enhances their ability to provide support and relief (WHO 2002d).

Beyond a fairly extensive study of traditional Chinese medicine, there has been little in the way of comprehensive attempts to examine traditional herbal medicines and identify new therapeutic opportunities that could be exploited in biomedicine. Some major pharmaceutical companies do from time to time examine the possibility of identifying new active substances on the basis of indigenous remedies, but it is not clear to what extent such efforts are productive compared with other approaches to innovative research.[11] Many substances of plant origin, once isolated and tested, prove to be similar to others already in use in a medicine, so that their further development is not justified. The effectiveness of traditional Chinese medicine (and many current indigenous systems) is based as much on a philosophical and cultural ethos as on specific natural ingredients. The efficacy of these treatments cannot be completely evaluated within the biomedical positivist model, yet no new methods to evaluate them have been developed that accommodate this holistic approach to medicines and therapies.

A large number of international organizations are committed to pressing the case for improved access to medicines

WHO has promoted a positive approach through its traditional medicines program, now closely associated with its main essential medicines activities. The "Global Strategy for Traditional and Complementary Medicine 2002–2005" (WHO 2002d) was released in 2002. The objective of the strategy is to discuss the role of traditional medicine in healthcare systems, the current challenges and opportunities, and WHO's role and strategy for traditional medicine. In a statement that accompanied the release of the document, WHO recognized that the use of traditional medicine is quite different from country to country and region to region. For example, in the African and Western Pacific regions, the member states consider that traditional medicine is a priority for healthcare in their regions, but in other parts of the world, traditional medicine is treated as complementary or alternative medicine (WHO 2002d).

Both patients and healers often have a clear understanding of the appropriateness of biomedical or traditional therapies for specific illness symptoms. The critical issue becomes the extent and quality of access to biomedical health services and supplies when traditional remedies do not exist to treat a specific disease such as AIDS or TB adequately. The belief that indigenous people will consistently first seek traditional modalities and the assumption that these patterns emerge from local (and immutable) systems of belief, rather than from economic and geographic exigencies, must be viewed with critical caution.

Unjustified assumptions about costs and the prospect of savings must be avoided. For example, herbal medicines prepared by patients or their families from the local vegetation may be available at negligible cost. However, the complexity of this issue is illustrated by the fact that people will often pay more for the services and medicines of a traditional healer than they would pay to seek biomedical treatment. This draws attention to the importance of understanding and trust between patients, their families, and practitioners. It also frequently shows the need for culturally appropriate communication and approaches to interventions.

Overview of major actors

A broad spectrum of specialties, interests, and levels of engagement have characterized international responses to access to medicine issues in recent decades. From issues of safety, quality, and efficacy to those of financing, infrastructure, and human resource needs, increasing attention has been paid to the complexities of getting the medicines from the manufacturer to the patient. Behind the scenes, the focus is on the R&D of new medicines and vaccines.

Today a large number of international organizations of various types are committed to pressing the case for improved access to medicines or working to bring about progress. There is a concern that these groups are in a sense competing with one another and with established donor agencies for funding. The emergent AIDS Medicines and Diagnostics Facility within WHO, if it develops as anticipated, may provide a model for coordinating efforts, as could

It is difficult to obtain a clear overview of the current levels of assistance for medicines because of a lack of coordination and transparency

the existing interagency coordination group for pharmaceutical issues (Boatin 2003; IAPCG 2000). The Global Drug Facility (GDF) was an initiative developed and implemented in 2001 by WHO, the Global Partnership to Stop TB, and the governments of 20 high-burden TB countries. The GDF, originally funded by the Canadian government, funds and facilitates movement of high-quality, generic anti-TB medicines into the countries most in need of them.

The CMH and agencies working on health in developing countries have been consistently recommending a drastic increase in the levels of aid provided for health in poor countries. The working group strongly agrees, but notes that it is difficult to obtain a clear overview of the current levels of assistance for medicines because of a lack of coordination and transparency among and between all of the major actors in this effort (CMH 2001; Troullier and others 2002).

Multilateral agencies

This section provides a brief history of efforts to improve access to medicines and some of the national and international institutions that began and developed these initiatives. The recent advent of public-private initiatives to address health and medicines challenges are also highlighted. Donation programs are considered in their own section. Although these actors should be congratulated for all they have done to bring medicines issues to the forefront in countries and in international forums, much remains to be done to make a substantive and lasting difference.

World Health Organization. Created in 1946, WHO limited its involvement in the area of medicines during its first quarter century to purely technical issues, particularly standards of manufacturing and quality control (Quick 2003). In the mid-1970s, then Director-General Dr. Halfdan Mahler introduced a broad change of policy in the organization as a whole, directing it toward improvement in basic health services and coverage; the concepts of "Health for All" and reliance on primary healthcare were introduced. To some extent this change in policy was a reaction to demands advanced by the growing membership of newly independent developing countries (Lunde 1984). In 1975, WHO introduced the Essential Drugs Concept as one of the basic components of primary care. Essential medicines were initially defined as "those considered to be of utmost importance and hence basic, indispensable, and necessary for the health needs of the population. They should be available at all times, in the proper dosage forms, to all segments of society" (WHO 1977). Expanded and revised, but not essentially different, definitions were adopted a quarter of a century later, in 2002 and 2003 (Laing and others 2003).[12]

Following the introduction of the concept of essential medicines, WHO went on to detail which specific medicines were safe, cost effective, and affordable, with particular consideration given to meeting the most pressing needs of the developing world. This approach led to the publication of the first model list

The WHO has become closely involved in measures to promote the appropriate use of medicines

of 224 essential medicines in 1977, largely based on similar lists developed in countries ranging from Papua New Guinea to Cuba to Mozambique. The WHO list was explicitly intended to serve only as a model for member states, and many national lists developed since that time have closely adhered to it. The model list has undergone regular review and revision, but it has only gradually expanded in size. The principal difference today is that a number of recently developed medicines, still under patent protection, have been included where they represent important innovations, notably in the treatment of AIDS. The list is now complemented by the existence of a model formulary providing basic prescribing information for all of the medicines listed on the model list (WHO 2003b).

The increasing need for medicines policies within developing countries by other UN organizations, notably the United Nations Industrial Development Organization (UNIDO), the United Nations Conference on Trade and Development (UNCTAD), and the United Nations Children's Fund (UNICEF), led WHO to create its own Action Program on Essential Drugs in February 1981. This expanded program forms the primary component of what is now the WHO Department of Essential Drugs and Medicines Policy.

The WHO has become closely involved in measures to promote the appropriate use of medicines (WHO 2002c) and has development standards in many fields, such as tools for developing national medicine policies (WHO 2002d); means of measuring medicine utilization and needs (WHO 1988, 2000f; Dukes 1993); and informative aids to critical procurement (MSH and WHO 2000). Training courses for national staffs are also provided in relevant areas. In these and other matters relating to pharmaceuticals, WHO currently works closely with a series of other global organizations, including UNICEF, UNCTAD, UNFPA, UNAIDS, and the World Bank, while an interagency coordination group for pharmaceutical issues has operated for several years (IAPCG 2000).

In 2001 the WHO, in collaboration with UNICEF, UNAIDS, and UNFPA and supported by the World Bank, started the Pilot Procurement Quality and Sourcing Project (known more commonly as the prequalification project or scheme[13]) for antimalarials, anti-TB medicines, antiretrovirals, and HIV and AIDS diagnostics. One of the most important achievements of this project was to standardize the criteria for assessing the pharmaceutical product dossiers and site inspections used by the participating UN agencies. Using a voluntary process, an increasing number of priority medicines for AIDS, TB, and malaria are now being assessed by the WHO and a team of international inspectors for their compliance with these standards. A list of prequalified products is published and updated every two months; some products have also been delisted on the basis of additional information. Although not intended for this purpose, the World Bank, the Global Fund to Fight AIDS, Tuberculosis and Malaria (GFATM), and many national procurement agencies are increasingly considering this information in their procurement decisions.

In 2003, the WHO announced the 3 by 5 Initiative to expand antiretroviral treatment for AIDS to 3 million people in developing countries by 2005

On World AIDS Day 2003, the WHO announced the 3 by 5 Initiative to expand antiretroviral treatment for AIDS to 3 million people in low- and middle-income countries by 2005.[14] As part of this effort, the WHO has established the AIDS Medicines and Diagnostics Facility, which aims to improve the supply of quality, effective HIV/AIDS diagnostics and medicines in developing countries. The World Bank and the GFATM increasingly rely on the WHO prequalification scheme or registration in Pharmaceutical Inspection Cooperation Scheme countries to identify good-quality medicines for AIDS, TB, and malaria.

World Bank. The World Bank has been significantly involved in medicines policy, procurement, and supply in low- and middle-income countries for many years. A pharmaceuticals group operates within the Health Systems Development component of the Health, Nutrition and Population department. According to a strategy paper on World Bank pharmaceutical policy and financing published in 2000, the focus of the bank's activities in medicines should be in the following areas: pharmaceutical sector reform; comprehensive national pharmaceutical sector assessments during project development in the health, nutrition, and population lending sector; support for greater transparency and accountability in pharmaceutical lending activities; lending that promotes policy and systems development and targets the poor; use of public-private partnerships to promote access to new essential medicines for the poor, including increased use of the private sector as a technical resource; and increased incentives for R&D aimed at diseases of the poor. The World Bank is a member of the Interagency Pharmaceutical Coordination Group, along with UNAIDS, UNFPA, UNICEF, and WHO. The Multi-Country AIDS Program (MAP) is the bank's long-term (12–15 years) program to scale up existing interventions to fight the AIDS pandemic. MAP projects include medicines procurement and supply. The bank recently published a very useful technical guide to HIV/AIDS medicines procurement (Taylor 2004). Among other relevant information, the guide promotes pro-generics procurement approaches and WHO prequalification, and offers important information for improving procurement, supply, and distribution of medicines in developing countries.

Global Fund to Fight AIDS, Tuberculosis, and Malaria. The GFATM was created as a pioneering financing mechanism for country-level efforts to combat AIDS, TB, and malaria. It is the largest health-related public-private partnership. It is an independent entity, governed by a board of directors that includes representatives from donors, the UN, civil society, and the private sector. Following an approved, focused application process, fund money can be used to purchase medicines to address any one or all of these three major diseases. The fund contributes to key access needs in two ways. First, it provides needed financing for medicines purchases that countries otherwise could not afford.

A number of bilateral agencies have been particularly important in funding related to medicines

Second, it has laid down very specific criteria for countries to meet concerning procurement, supply, and distribution of medicines to help ensure that quality medicines are bought and that those medicines get to the people who need them. It is hoped that these requirements will serve as a catalyst for governments to improve their current regulatory, procurement, supply, and distribution systems, all of which are key to improving access to medicines for the entire health system. A major problem, as noted by many public interest nongovernmental organizations (NGOs), recipient governments, and the fund management itself, is the continued failure to receive adequate donor financing, sometimes despite donor promises for funds.

Bilateral donors

All donor support to the health sectors of developing countries is necessary and appreciated. However, current funding is falling far short of what is needed. For example, Médecins Sans Frontières (MSF) estimated in 2002 that donors had pledged just 8 percent of the estimated funding needed to scale up the global response to AIDS. As mentioned above, the GFATM remains seriously underfunded. The WHO 3 by 5 Initiative is not adequately resourced. According to the UN Millennium Project Working Group on Malaria, the estimated funding needs to deliver antimalarial combination therapies is $300–500 million per year in Sub-Saharan Africa alone.

Within the bilateral donor community, a number of agencies have been particularly important in funding related to medicines, either by the size of their funding allocations, their prioritization of medicines issues and provision, or both. All bilateral donors that have significant commitments to health development strongly support the development and launch of effective AIDS treatment plans in all countries that need them. For example:

- The Danish International Development Agency (Danida) has a long history of supporting the establishment of essential medicines programs and strengthening medicines regulatory and supply systems. This longstanding commitment is under pressure, though. Since 2001, the government has been adopting a much more conservative approach to development aid.
- The U.K. Department for International Development (DFID) has a well-rounded aid program for medicines. It has just launched a new major commitment to increasing access to essential medicines (DFID 2004b).It is the second largest bilateral contributor to the GFATM. DFID money can be used to purchase medicines in national programs.
- The Dutch Directorate General for Development Cooperation (DGIS) is also a large contributor to international and national organizations, including WHO and many NGOs, for activities related to medicines. Over the last 15 years, it has been the largest cumulative contributor to the WHO Department of Essential Drugs and Medicines Policy. DGIS

Foundations have played leading roles in public-private initiatives for medicines and vaccines development

been a major supporter of efforts to improve the quality of medicines information, including medicines promotion and improving methodologies for collecting and analyzing medicines prices.

- The Swedish International Development Cooperation Agency (Sida) has provided similar health systems support in Southeast Asian countries, such as Lao People's Democratic Republic and Viet Nam, as well as selected countries in Africa, such as the setting up of the first essential medicines program in Angola in the late 1990s.
- The U.S. Agency for International Development (USAID) will certainly have an impact on access to AIDS treatment in the 15 developing countries it selected to receive funding from the $15 billion President's Emergency Plan for AIDS Relief (PEPFAR). Already the plan has sparked tremendous controversy by initially forbidding PEPFAR money to be used to purchase generic fixed-dosed combination antiretrovirals. In response, the U.S. government has initiated a special U.S. Food and Drug Administration approval process for generic medicines produced outside the United States. This move was adopted by the United States in lieu of the WHO prequalification project for antiretrovirals, TB medicines, and antimalarials. WHO prequalification standards are stringent and include regular quality monitoring, as well as prompt notification when a medicine is removed from the prequalification list.

Private foundations

Foundations that support health development also play a very important role in health funding. In particular, foundations have played leading roles in public-private initiatives for medicines and vaccines development. A few examples relevant to a discussion of access to medicines include the following organizations:

- The Bill and Melinda Gates Foundation has health as a major focal area in its strategic funding efforts. The Gates Foundation promotes the public-private initiative model. It has made large contributions to the GFATM, as well as to initiatives to promote the development of new medicines and vaccines. The foundation provides funding to the WHO 3 by 5 Initiative. It is a major funder of the Strategies for Enhancing Access to Medicines program at the U.S.-based consulting firm, Management Sciences for Health (MSH), which supports local public-private initiatives and public health–oriented private sector interventions.
- The William J. Clinton Foundation was responsible for a breakthrough in antiretroviral benchmark pricing in 2003. The foundation was able to secure agreements from manufacturers to supply selected countries with antiretrovirals for $140 per person per year (which is roughly half of the previous lowest price for treatment in developing countries). However, the arrangement has stringent conditions, including a three-year,

Because of severe problems of affordability in poorer countries, the role of nonprofit wholesale suppliers is very important

noncancellable contract; demonstration of purchaser financing to pay for supplies over the three years; and very high purchase volumes with each order. The arrangement demonstrates the important role that ensured demand and payment can play in setting delivery prices. The extent to which purchasers, often developing country governments that rely on donor financing that is not administered in a manner amenable to such arrangements, can avail themselves of these agreements and at what scale remains to be seen.

- The Rockefeller Foundation for several years has provided support to promote research and discussion in the area of access to essential medicines. It been a key player in the product development initiatives of the Global Alliance for TB Drug Development (GATB) and the Medicines for Malaria Venture (MMV). It has funded a policy series initiative on access to affordable medicines, bringing together leading experts for policy discussions in workshops at its Bellagio center. It is one of the funders of the WHO–Health Action International (HAI) pricing survey methodology initiative.

Supply organizations

Because of the severe problem of affordability in poorer countries, the role being played by nonprofit wholesale suppliers is very important. Supplies are obtained from low-cost sources, including research-based pharmaceutical manufacturers, and are distributed to developing countries either directly or through development finance programs.

- The International Dispensary Association (IDA), based in the Netherlands, has been supplying quality (mainly generic) medicines to governments and other public or nonprofit sector clients in developing countries for more than 30 years. Its routine publication of drug price lists has increased market transparency to the benefit of procurement specialists seeking the best return on their (often limited) budgets.
- Missionpharm in Denmark primarily supplies missionary health organizations.
- UNICEF Supply Division maintains a global warehousing and supply system to deliver medicines to developing countries at favorable prices. They engage in direct negotiations with manufactures and suppliers to secure patented or generic medicines at highly favorable prices to developing countries.

Emergence of a global access to medicines advocacy campaign by nongovernmental organizations

No single NGO can be credited with having first recognized and tackled the size of the access to medicines problem. The Consumer Project on Technology (CPTech), a public interest group founded by Ralph Nader in the United

No single NGO can be credited with having first recognized and tackled the access to medicines problem

States, was one of the first NGOs to recognize that TRIPS was going to have a negative impact on public health. Soon after TRIPS came into effect in January 1995, it launched a project focusing on TRIPS and healthcare, with specific concerns about the impact of TRIPS on medicines prices.

By 1998, HAI, a public interest network working exclusively on medicines issues, became actively engaged on a regular basis with the issues of intellectual property and access to essential medicines. HAI and CPTech were influential in pressuring the WHO to address the impact of TRIPS on access to medicines in the World Health Assembly in May 1998. The intense fight over language about trade and medicines in the proposed WHO Revised Drug Strategy foreshadowed what was to happen with intellectual property, the WTO, and access to medicines. The outcome eventually allowed the WHO to assume a more central role in gathering information and informing public debate about access to medicines and intellectual property issues.

By 1999, MSF, having just won the Nobel Peace Prize and deciding to devote the prize money to finance a global access to medicines campaign, became the de facto leader in the access movement, joining CPTech and HAI to organize a large conference on access issues just prior to the WTO ministerial meeting in Seattle. In 2000, Oxfam also launched a major access to medicines campaign, and Third World Network played a key role in developing technical assistance to developing countries on how to formulate pro-public health intellectual property legislation as they become TRIPS compliant. The Kenya Coalition for Access to Essential Medicines was successful in getting pro-public health TRIPS safeguards into Kenyan intellectual property legislation in 2001.

Linking among NGO groups working on access to medicines issues, especially those focused on access to affordable antiretrovirals in developing countries, was spurred by a lawsuit by 39 pharmaceutical companies against the South African government to prevent it from enacting a new piece of medicines legislation that permitted parallel importation of medicines (parallel importation is a flexibility permitted under WTO rules). Eventually, the pharmaceutical companies withdrew their complaint, due in no small part to the impact of a global and local advocacy campaign against their position. Treatment access groups in developing countries, for example the South African Treatment Access Campaign, have become effective advocates to pressure governments to increase AIDS treatment and provide affordable medicines. A Pan-African Treatment Access Movement has been launched. MSF continues to produce valuable information for policy debates at the international level and works to increase access to medicines through its many developing country-level programs.

It is clear that the emergence of a strong and ongoing global advocacy NGO network on access to medicines, especially antiretrovirals, has been a crucial boost to everyone working to increase access to medicines in developing countries. The impact of the network has been felt especially at the international

The emergence of a strong and ongoing global advocacy NGO network on access to medicines has given a crucial boost to work on access to medicines

policy level in new ways. For example, the substantive participation of a public interest civil society voice has influenced WTO actions on TRIPS and medicines (such as the Doha Declaration on TRIPS and Public Health [WTO 2001]); helped prompt two major international commissions on intellectual property (the U.K. Commission on Intellectual Property Rights and the current WHO Commission on Intellectual Property Rights, Innovation, and Public Health); helped support an increased role for the WHO on medicines issues; and helped influence legislation and treatment policies at the country level. Strong support for generics competition, independent monitoring of medicines prices, and other valuable supply information has helped lower medicines prices, particularly for antiretrovirals. Probably most importantly, especially at the country level, has been the way in which access to medicines issues have catalyzed and empowered citizens to engage in the political and policy process in new and important ways within the health sector.

Many well established NGOs provide essential health service delivery in developing countries. Some are also involved in advocacy on health and medicines issues. However, many are not. It is important to acknowledge, then, the vital role such organizations as the Red Cross and Red Crescent, CARE, and the many ecumenical organizations such as national Christian health associations, the Medical Mission Sisters, Catholic Relief Services, and others play in providing healthcare in developing countries.

Pharmaceutical industries

Pharmaceutical industries are key actors in ensuring the availability of needed medicines. The sector is not homogenous. It can be divided most easily into two main categories: the generics industry and the research-based industry. It should be kept in mind that even within either of those categories, the companies involved vary significantly in how they respond to medicines needs in developing countries.

It is beyond the scope of this report to provide comprehensive overviews of each company. The focus here will be on the provision of medicines in developing countries and the respective roles of the industries and the issues surrounding those roles. Notably, the role of the pharmaceutical industry is probably the most contentious and complicated of any in the discussions surrounding access to medicines for the poor. The debate about the pharmaceutical industry is very often very charged and misrepresentations are made by both critics and supporters.

The challenge for this working group has been to address the pharmaceutical industry's crucial role in a way that can be balanced and yet still call on all major actors to rise to the undeniable requirement to change the way they respond to the need for medicines in developing countries. It is very likely that the scale of need will demand fundamental rethinking of how medicines and vaccines are developed and marketed. Expectations and new approaches should keep in

The generics industry is the largest supplier of medicines to the developing world

mind the fundamental nature of these industry actors: they are in the medicines business to make a profit. The extent to which pressures and realities will cause a shift to include a socially responsible component remains to be seen.

Generics pharmaceutical industry. The generics industry is, in unit terms, the largest supplier of medicines to the developing world and to countries in transition. According to WHO, in low- and middle-income countries in 2000, the generics market share was at least 50 percent; the largest low-income country generics market share was Bangladesh at 70.9 percent. It is generics manufacturers that have been responsible for ensuring the wide availability of affordable generic versions of essential medicines in developing countries. Notably, more than 90 percent of medicines listed on the WHO essential medicines list are off patent.

Historically, generics producers have concentrated on replicating originator products once they were off patent. Innovation was very limited for many reasons. However, the generics sector is changing, and more R&D is being undertaken. For example, Zithromax®, a key antibiotic now licensed to Pfizer, was discovered and developed by a Croatian generics firm that was working with a U.S. subsidiary. The National Association for Rare Diseases produced the first medicines under the U.S. Orphan Drug Act. Generics firms have developed combination antimalarials.

Many generics firms concentrate on the large-scale production of medicines at low cost, and their prices, under international nonproprietary names, are commonly only a small fraction of those charged by originator companies (International Drug Price Indicator Guide 2003).[15, 16] In the case of antiretrovirals, price competition from generics has been responsible for pressuring originator medicines prices downward, to compete comparably with them in some cases. Low generics prices are low partially because these firms do not bear R&D costs. They also reflect the fact that unit profits are usually lower and that these firms do not, as a rule, engage in advertising or promotion for their products. Furthermore, production is, to a large extent, undertaken in countries where manufacturing overhead costs are also lower.

Major generics manufacturers are quite similar in their staffing, facilities, and product quality to the staffing, facilities, and product quality of research-based companies. In fact, the largest generics manufacturer in the world, Sandoz, is a subsidiary of the giant research-based transnational company, Novartis. The best firms adhere fully to internationally recognized standards of good manufacturing practice (GMP). Some are based in industrialized countries. A number actually choose to provide their products with brand names.

Quality generics are fully equivalent to originator versions in terms of quality, safety, and efficacy and are usually less expensive. Again, it is important to stress that neither the generics nor the research-based industries are monolithic. In addition to the internationally competitive generics manufacturers

Without generics competition, the availability and affordability of key essential medicines in developing countries would suffer

there are at least two other kinds of production mechanisms that exist within the context of developing countries. In middle-income countries, such as Brazil, state-controlled generics firms produce medicines to meet national requirements. And, in low-income countries, such as Kenya or Zimbabwe, local generics manufacturers may supply the local market with essential medicines, with some form of preferential government subsidy.[17]

In recent years, research-based and generics manufacturers have pursued voluntary licensing agreements for key medicines. This avenue is one way for local firms in developing countries to produce quality medicines in cooperation with the license holder. The arrangements for this production should include technology transfer and capacity building of local staff to be sustainable.

In some cases, generics producers in low-income countries do not comply with GMP, limiting themselves to simple, high-profit formulations. These firms do not want to have to meet bioequivalency requirements, which are required for WHO prequalification for medicines to treat AIDS, TB, and malaria. This approach limits the contribution of local generics manufacturers in developing countries to compete in the markets for these essential medicines. Producers wishing to ensure quality standards can avail themselves of WHO technical assistance to meet GMP and prequalification requirements. They can also pursue cooperative agreements with larger manufacturers that can provide technology transfer and staff training.

Key achievements for the generics industry include taking the lead in providing price competition for antiretrovirals that prompted real reductions in the price of antiretrovirals offered from originator manufacturers. Before increased generic competition on antiretrovirals, originator companies set price discounts as they saw fit. While voluntarily discounted originator prices, such as those under the joint UN–Industry Accelerated Access Initiative, did lower costs from $12,000 to $2,000 per person per year in selected developing countries, the costs were still unaffordable for poor governments and for the majority of the population in need. It took generics competition to bring the price below $1,000 and eventually to $250–$300. It is clear that without generics competition and their robust involvement in developing country markets, the availability and affordability of key essential medicines would suffer.

Research-based pharmaceutical industry. The research-based industry is the dominant supplier of medicines to industrialized countries. The most significant contribution the research-based industry can make to the goal of expanding access to essential medicines is to discover and develop new medicines. Most of these companies are also fully committed to working in partnership with others to ensure that these medicines are then available as broadly as possible. Private research-based pharmaceutical companies have produced the overwhelming majority of new medicines now on the market. For example, three-quarters of medicines now on the WHO essential medicines list,

The development of medicines that target the health problems in developing countries requires innovative approaches

although discovered in the public sector, were brought to market successfully by research-based companies.

The principal characteristic of these companies is the presence of innovative research. Much of this creative work is undertaken in-house, but these companies have also been increasingly successful in drawing on the output of publicly funded academic and institutional research. In fact, research-based industry recognizes the important relationship it shares with academia and public sector research agencies. Although scientists in research-based pharmaceutical and biotechnology companies contribute significantly to basic research, it is also true that publicly funded investigators have traditionally conducted the bulk of basic biological research. However, the research-based pharmaceutical industry continues to lead in the more applied research activity that ultimately results in the discovery and development of most new medicines.

The research-based industry is headquartered almost exclusively in industrialized countries and is increasingly engaged in a series of large and generally multinational enterprises. It is highly profitable: in the United States, the pharmaceutical industry has been by far the most profitable corporate sector over the past 30 years. Although it is an industry engaged in providing vital public goods, it remains driven by its corporate mandate to maximize profits. Therefore it is not surprising that its focus is on developing products for affluent markets. As a result, the research-based industry has come under criticism for not responding adequately to medicines needs of the poor in developing countries.

However, indications are growing that research-based companies, by and large, are doing more and wish to do still more. To promote the development of medicines that target the health problems in developing countries, innovative research, development, and financing arrangements are needed. Some examples of this include recent collaborations between the WHO, industry, and other stakeholders to lower the prices of key medicines, such as antiretrovirals, and those for treatment of MDR-TB. Originator companies have also expanded public-private alliances and interactions, donation programs, and technical assistance to help fill the capacity gap. A few examples include the African Comprehensive HIV/AIDS Partnership in Botswana, where 26,000 patients receive antiretrovirals donated by Merck; Uganda's first Infectious Disease Institute built by Pfizer in Kampala to provide a training site for physicians and technicians and a laboratory for monitoring and testing patients undergoing AIDS treatment; AIDS programs initiated by Bristol-Myers Squibb in five Southern African countries that provide antiretroviral medicines, as well as inputs into health management, medical research and education, community education and outreach, and capacity-building programs for women and children. The Accelerated Access Initiative is a cooperative endeavor of seven research-based pharmaceutical companies, UNAIDS, WHO, UNICEF, the

Public-private initiatives hold promise as a way for pharmaceutical industries to contribute positively on a variety of important fronts

UNFPA, and the World Bank to negotiate price discounts for antiretrovirals and provide treatment in selected developing countries.

The research-based pharmaceutical industry has indispensable scientific expertise to help solve the problem of access to essential medicines. However, it cannot do this alone. Leadership from the global public sector (including donors, multilateral organizations, and academia) will be required to create the necessary conditions to develop medicines for priority health conditions in low-income countries.

Public-private initiatives

One of the key challenges in achieving any development goal is to effectively harness the contributions of major actors. Meeting the health needs of developing countries is daunting and complex and requires finding new ways for major actors to work together. The pharmaceutical industries have a vital role to play and resources to contribute. However, the conundrum has often been finding a framework that allows disparate public and private actors to work toward common goals successfully. The rise of public-private initiatives in recent years is a trend that holds considerable promise as a way for pharmaceutical industries to contribute positively on a variety of important fronts. Since these types of interactions are diverse and new, it will be important for all of the actors to monitor and learn from initial experiences. Initial experiences point up much hope for the evolution of public-private initiatives as an effective means of collaboration, as well as pointing out inevitable limitations to be managed.

It is important to note that the term *public-private initiatives* is not entirely straightforward (box 1.2). In addition to coming in varying shapes, sizes, and modes of operation, public-private initiatives are being created to address a range of public health needs. Global interactions relevant to medicines can be divided into four main categories:

- Public-private product development interactions, (such as GATB, MMV).
- Donations (such as Mectizan®).
- Public-private coordination or financing mechanisms (such as Stop TB and the GFATM).
- Public-private interactions for strengthening health systems (such as the African Comprehensive HIV/AIDS Partnership in Botswana).

Public-private initiatives typically comprise various combinations of international and multilateral organizations (such as WHO, World Bank, and UNICEF), philanthropic foundations, research-based pharmaceutical companies, academic institutions, bilateral donors, and NGOs. These collaborative efforts, when successful, deliver the following positive outcomes:

- Deliver critical funding.
- Draw attention to health threats that may not be widely known.
- Share knowledge and resources.
- Build the numbers needed to facilitate volume-related discounts.

Box 1.2
What are public-private partnerships?

Source: Richter 2004.

The literature on global public-private partnerships is confusing for at least two reasons. First, even though public-private partnerships have been promoted for several years, there is no one single, agreed-upon definition. Second, most discussions do not distinguish between public-private partnerships as a policy model and actual examples of public-private partnerships and interactions that have been undertaken. According to the 2003 report of the UN Secretary-General on "Enhanced Cooperation between the United Nations and All Relevant Partners, in Particular the Private Sector," partnerships are commonly defined as voluntary and collaborative relationships between various parties, both state and nonstate, in which all participants agree to work together to achieve a common purpose or undertake a specific task and to share risks, responsibilities, resources, competencies, and benefits.

Many of the relationships that are currently called partnerships are not in fact new. Researcher Ann Zammit, who reviewed UN-business partnerships extensively for the United Nations Research Institute for Social Development, remarks in *Development at Risk,* "The term [partnership] covers a multitude of activities and relationships, perhaps best conceptualised as a special case of 'close' rather than 'arms-length' relationship between government and business" (Zammit 2003, p. 5).

The three levels of analysis about public-private partnerships are often blurred because the term "public-private partnership" is used to describe

- A policy paradigm (including its underlying framework of thought and ideology).
- Various categories of public-private partnerships or interactions, such as donations of pharmaceuticals or legally independent global health alliances.
- A specific public-private partnership or interaction, such as the Malarone® Donation Programme or the Global Alliance for Vaccines and Immunization (GAVI).

It is critical to make a theoretical distinction between these three uses of the term.

- Achieve specific public health disease management objectives (such as the Global Polio Eradication Initiative).

The private sector, by its very nature, seeks interactions and investments that deliver a clear and substantial return, financially and/or through branding or expanded influence in policy spheres or markets. In theory, the private, for-profit sector can and should assist the public sector in identifying ways that progress toward the essential goal of access to medicines can benefit from private sector, competitive initiatives. Room exists to increase the pharmaceutical industries' contributions to solving the problem of access to medicines. For example, using a joint venture model as the organizing framework for an alliance permits some formal separation of the for-profit and not-for-profit activities. This is indeed the case in many agreements negotiated by public and private institutions—in particular in the area of developing new medicines and vaccines.

A 2002 report by the Bill and Melinda Gates Foundation illustrates the application of a business model in the development and operation of global health alliances (Bill and Melinda Gates Foundation 2002). The report notes that more than 80 percent of the past and present alliances that were reviewed were deemed to be successful. The authors lay out an operational framework

A public-private initiative approach appears to be particularly effective in the area of new medicines and vaccines development

for analyzing an alliance's effectiveness that includes the existence of clear and compelling goals and clear scope (as defined by geography, patient populations, functional activities, and time). Other elements such as the nature of alliance structure, agreement on success metrics, milestones, partner contributions, and sufficient number of staff are discussed (Bill and Melinda Gates Foundation 2002).

Given all of the challenges facing the world community to reduce poverty and disease in developing countries, it is very welcome that the private sector's expertise and resources can be harnessed in new ways. Research-based industry is embracing the public-private initiative approach as a way to increase their contributions substantially in programs targeted at the poor in developing countries. The Partnership for Quality Medical Donations reports that its partners contributed $1.4 billion in donated medicines in 2003. It is clear that long-running large-scale public-private initiatives for the elimination of priority diseases of poverty, which involve sustained medicines donations and collaboration with partners in management, training, and implementation, have been successful. Examples are the Merck Mectizan® donation program (for onchocerciasis and lymphatic filariasis, the latter in conjunction with GlaxoSmithKline's donation of albendazole), Pfizer's donation of Zithromax® in the Trachoma Initiative, and Boehringer Ingelheim's now open-ended donations of Viramune® for the prevention of mother-to-child transmission of HIV.

A public-private initiative approach also appears to be particularly effective in the area of new medicines and vaccines development. This is especially true for the diseases strongly associated with the consequences of poverty, including AIDS, malaria, TB, sleeping sickness, Chagas disease, leishmaniasis, Buruli ulcer, dengue fever, leprosy, lymphatic filariasis, and schistosomiasis. Below is a list of major initiatives that focus on developing new medicines or vaccines, all of which include multiple public and private actors working often in innovative ways together within the interaction:

- *Aeras Global TB Vaccine Foundation.* Aeras was founded in 1997. Since 1999 it has focused exclusively on the development of an effective TB vaccine. Formerly the Sequella Global TB Foundation, Aeras has worked closely with the Bill and Melinda Gates Foundation.
- *Drugs for Neglected Diseases initiative (DNDi).* Conceived and organized by MSF, members include research institutes in Brazil, India, and Kenya, the Malaysian Ministry of Health, and the Pasteur Institute. The WHO Special Programme for Research and Training in Tropical Diseases (TDR) is an observer. The initiative aims to develop medicines to combat neglected and most neglected diseases. Current work includes the development of two fixed-dose combination artesunate antimalarials and seven projects to develop medicines against visceral leishmaniasis, Chagas disease, and trypanosomiasis.

It is important to monitor public-private initiatives, to measure their effectiveness and learn from experience

- *Global Alliance for TB Drug Development (GATB).* GATB was formed to accelerate discovery and ensure that there are better and more affordable agents to fight TB. The alliance makes support to institutions in TB-endemic countries a priority. It works with research-based companies to identify promising substances and move them along the development pipeline.
- *International AIDS Vaccine Initiative (IAVI).* The main aim of this initiative is to support and find ways to speed up discovery and development of an effective HIV vaccine. IAVI receives major financial support from a number of major philanthropic foundations, the World Bank, and nine national governments.
- *Medicines for Malaria Venture (MMV).* MMV selects, guides, funds, and develops research done by others and works in partnership with research institutions, ministries of health, disease control programs, research-based industry, academia, and NGOs to improve the availability of safe, effective, and affordable antimalarials. Two related initiatives include the Malaria Vaccine Initiative and the European Malaria Vaccine Initiative.

Without going into detail on all product development public-private initiatives, it is worth noting the following initiatives and projects to illustrate the range of activities under way: the South African AIDS Vaccine Initiative, the International Partnership for Microbicides, the Global Microbicide Project, the Microbicide Development Project (for anti-HIV microbicides), the Institute for OneWorld Health, the Pediatric Dengue [Fever] Vaccine Initiative, the Human Hookworm Vaccine Initiative, the Rotavirus Vaccine Accelerated Development and Introduction Plan, the Pneumococcal Vaccine Accelerated Development and Introduction Plan, and the Consortium for Industrial Collaboration in Contraceptive Research.

Global public-private coordination and financing mechanisms include the Stop TB Partnership, the Roll Back Malaria Partnership, the Global Alliance for Vaccines and Immunization (GAVI), and the GFATM.

As already pointed out, public-private initiatives, with the exception of a few long-running donations programs, are relatively new and diverse. It will be important to monitor the various types, measure their effectiveness, and learn from experiences. Monitoring and further operational research should cover the more objective and quantitative aspects mentioned above as well as the following process issues:

- *Sustainability.* Public-private initiatives should have timelines and funding commitments tied to reaching public health objectives, which may take many years to achieve. They may, therefore, need to be very long term or even open ended in initial phases. Short time horizons, for example, within GAVI or for GFATM project funding (five years, with no guarantees of further funding), are often counterproductive. Even if

With the global scale of recent health sector initiatives, transparency and decision-making processes are critical

actual allocations are yearly and subject to regular review and evaluation, sustainability is enhanced when resource commitments are known and driven by health needs.

- *Power inequities.* Inherent in many partnerships or interactions is that the developing country partners have no power to insist on the continued support and participation of either the private-sector partners or the international institutions.
- *Pro-poor focus.* Public-private initiatives, which are often very complicated to manage and implement successfully, may tend to default to commitments that are the most manageable and likely to succeed rather than those that reach the greatest number in the greatest need. Principles of equity and setting as a priority reaching the poor should be at the forefront of planning, implementation, and budgeting.
- *Competing initiatives.* Multiple initiatives in a single country or region will rely on the same (often overwhelmed) health system infrastructure to identify and diagnose patients and to deliver medicines.
- *Governance and accountability.* With the global scale of recent health sector initiatives, issues of transparency and decisionmaking processes have become critical. Democratic process is not the rule, and decisionmaking often is controlled by the multilateral institutions and the private-sector participants. GAVI is one example of an interaction that has come under scrutiny on these issues (Hardon 2001). Public-private initiatives need to be accountable to intended beneficiaries as well as the public-private initiative managers and funders. At the national level, government representatives should play leading roles in decisionmaking and accountability. These roles can be difficult when the government involved is relatively weak and dependent in relation to other participants.

Overcoming barriers to access

The Working Group on Access to Essential Medicines has organized its analysis and recommendations into three main categories: availability, affordability, and appropriateness.[1] One or more of these elements will be missing or deficient in settings where access to medicines is inadequate (Liu 2003). The WHO includes a fourth category: reliable health and medicine supply systems. The working group addresses medicine supply systems in the discussions of availability and affordability and recognizes the importance of the larger healthcare system as the context in which medicines are prescribed and dispensed.

The degree of confidence that people invest into a local health system is determined largely by the ability of the system to respond in a timely and consistent manner to the needs of patients. This in turn will often be directly related to the existence of a sustainable supply of essential medicines. Critical shortages of trained personnel and severely deficient health system infrastructures will also impede the safe and effective prescription and movement of medicines that are available. This interdependence demonstrates the considerable and often dialectical interaction between all of the various components that will ultimately determine access to essential medicines in any given location.

Innovation

Treating priority diseases of the poor is greatly hindered by a fundamental problem: the medicines required for many of the diseases and illnesses most prevalent in developing countries do not exist because of a lack of therapeutic innovation (MSF 2001). Byström and Einarsson (2001) estimated that, between 1975 and 1997, only 13 of 1,223 new chemical entities found to have useful pharmacological properties were for treating diseases predominantly prevalent in poor countries. Similar estimates have been made by others (Yamey 2002; Troullier and others 2002).

A reorientation of medicines research is necessary to make it better attuned to the needs of the poor

Another critical need is for new medicines to supplement or replace those to which microorganisms have become resistant, as is notably the case for malaria and TB.[2] A complementary approach to the resistance problem is to develop and employ new combinations of medicines. These combinations require proper investigation before they can be accepted and used, however, and again there is little commercial incentive to study them.

The lack of innovation in medicines needed to meet public health problems in poor populations reflects the fact that industrial research has been directed primarily toward treatments for diseases of industrialized countries, where chronic diseases associated with longer life spans (such as cardiovascular diseases and cancer) are the main causes of mortality and morbidity (Troullier and others 2002). While the *World Health Report 2003* (WHO 2003f) notes that chronic diseases are also becoming more prevalent in developing countries, the numbers of people affected are still far lower than the numbers affected by the major infectious diseases.

Despite substantial achievements, a dramatic ongoing need continues for new and innovative medicines and vaccines to fight current, emerging, and evolving health challenges. The need is likely to grow, and concerns exist that existing medicines, such as antibiotics, may be threatened by resistance before effective new medicines can be developed and that the incentives for vaccine development are failing. The research-based industry has started to respond, for example, by launching dedicated research facilities focused on TB, malaria, dengue fever, and other parasitic diseases. It would appear that public-private initiatives for the development of new medicines and vaccines may be a productive way of promoting innovative research, development, and financing within the current system.

Three of the most neglected diseases—African trypanosomiasis, Chagas disease, and leishmaniasis—for example, are now beginning to receive additional attention, in particular through new public-private initiatives. For African trypanosomiasis, an initiative by the WHO and three pharmaceutical companies—Aventis, Bayer, and Bristol-Myers Squibb—has been established.[3] There are several products available for leishmaniasis, developed by pharmaceutical companies working with the Special Programme for Research and Training in Tropical Diseases (TDR), a joint effort of UNICEF, UNDP, the World Bank, and the WHO. For Chagas disease, Roche has donated rights and technology to manufacture benzonidazole (the most effective medicine for this disease) to the Brazilian government. The only significant tropical disease for which there is no existing medicine is dengue fever. But even for this disease, five compounds are currently in stages of discovery and preclinical development, two are in Phase 1 trials, and one is in Phase 2 trials (IFPMA 2004).

A reorientation of medicines research is necessary to make it better attuned to the needs of the poor. This will require creative new research, development, and financing mechanisms. The for-profit private sector is not going to take

Public-private initiatives offer useful new models for organizing R&D among major actors

up needed innovation for major infectious diseases in poor countries without major involvement and subsidy from the public sector and an appropriate and supportive policy environment. Provided this environment can be realized, it is hoped that the research-based industry would also be willing to increase their regular R&D budgets that are devoted to priority diseases of developing countries.

A rethinking of public and private R&D investments should include an analysis of the role of academia. Although a great many fundamental discoveries potentially relevant to therapeutics emerge from academic research, academia does not itself have the structure to carry these through to development and marketing. Many academic units, with restricted budgets of their own, cooperate closely with pharmaceutical companies and receive industry funding for research projects. To a large extent, though, those research priorities are determined primarily by the industrial partner, which has an eye toward potentially profitable markets. To date, much of the discussion has centered too much on the role of industrial country–based academia in research. The substantive involvement of research institutions and scientists in developing countries should become the norm in all stages of R&D.

Public-private initiatives for new medicines and vaccines, such as MMV, IAVI, DNDi, and GATB, appear to be offering useful new models for how R&D can be organized among major actors. They are a main way to increase the involvement of researchers and institutions based in developing countries and to establish and strengthen ties among industry, academia, and others to combat major diseases of poverty at regional and global levels. Well organized and managed public-private initiatives help maximize the comparative advantages of all of the participants. Since most public-private initiatives are relatively new, it will be important to monitor their progress and outcomes.

Meeting R&D needs for the poor means taking risks by the public and private sectors and—importantly—requires creative new thinking by major actors. For example, innovative, open-source arrangements for sharing knowledge should be investigated and developed as justified to meet R&D objectives. Proposals to reorient the way the financial burden is shared among countries deserve serious analysis and discussion. Another proposal is to promote increased public financing for biotech firms or academic units working on the medicines, vaccines, and technology needs of the poor. Yet another idea for creating research incentives is to establish international drug or vaccine funds that will be used to purchase guaranteed quantities of medicines for use in developing countries. Recently, the Bill and Melinda Gates Foundation offered financial rewards for the discovery of medicines for neglected diseases. That foundation has also provided a major grant to the first nonprofit pharmaceutical firm that has a stated mandate to develop medicines for diseases of poverty.

The principle of profit-based and patent-buttressed incentives to enterprise and innovation is widely accepted. The principal drawback for medicines

New approaches require a financing structure to provide incentives to all major actors—governments, international organizations, academia, and industry

innovation of that approach is that 90 percent of this money still is spent on the health priorities of 10 percent of the world's population, according to the Global Forum for Health Research (GFHR 2004).

For new approaches to succeed, an adequate financing structure needs to be available that provides incentives to all major actors—governments, international organizations, academia, and industry. They all need to release themselves from longstanding notions about their roles and contributions and be willing to build trust and try new ways of working together.

New approaches should promote fresh strategic analysis, emphasizing a substantive role for representatives from affected countries. They should ensure technology transfer, development of research capacity, research leadership, and creation of manufacturing facilities in affected countries (Folb 2004).

The working group recognizes that successful innovation to help meet the Goals will require greater cooperation among all sectors (such as public and private sectors, academia, foundations, and the United Nations), substantially more financing from multiple sources, clearly setting priorities for research efforts, effective management, and promoting technology and knowledge transfer. WHO should take a leading role in promoting R&D that meets the public health priorities of developing countries. For instance, the WHO Commission on Intellectual Property Rights, Innovation, and Public Health should examine alternative international models to the current patent-based system for priority setting and financing of health R&D. Drug regulatory process reforms and harmonization—for example, the International Conference on Harmonization (ICH)—need to better reflect and serve the needs of developing countries. Traditional knowledge and medicines continue to be marginalized, to the detriment of consumers. Vigilance surrounding all aspects of pharmacological practice in developing countries needs to be strengthened.

The answers all point to the need for considerable change. The working group appreciates that the public must continue to support research-based firms. It would seem no less than equitable that these innovation costs should be borne primarily by the nations with the broadest shoulders, such as heavily industrialized countries with strong economies that are capable of sustaining relatively high prices for the medicines that they require.[4]

Production

Even medicines that have been developed may not be in production if they are considered unprofitable or if supply chain requirements are not sufficiently known. Some medicines developed virtually to the point of application are abandoned; others are discontinued because of disappointing sales. At present very few low- or middle-income countries are able to compensate for the unavailability of particular medicine substances on the world market by producing them at the national level.[5]

Generics producers in low- and middle-income countries have been helping to ensure price competition for key essential medicines

Local production in such countries is rarely at a sufficient economic scale and technological level to take up such a challenge, as it is usually limited to processing imported active substances into finished form. From an economic perspective, medicine production requires considerable capital investment, adequate technical staff, and basic infrastructure. The cost of quality assurance systems is also considerable. All of this should be seen in the perspective of fierce global competition. Local production will usually be economically feasible and sustainable only above a certain turnover. More than a decade ago, consultations within the World Bank led to the view that, with certain well defined exceptions, new investment in medicine production in countries at a low level of development was unlikely to be justified. Quite apart from the difficulties in maintaining adequate staffing and ensuring sound quality standards, the products of this relatively small-scale production would be unlikely to compete in cost with generic products produced on a large scale in countries such as India and China.

But the political aspect of local production also needs to be considered. For example, Brazil was able to negotiate reduced prices of antiretroviral medicines by threatening compulsory licensing, because national production capacity was actually available. National production capacity also allows for voluntary licensing, as was recently the case in South Africa and in Kenya. However, voluntary licensing, especially if it is achieved under the threat of compulsory licensing, does not automatically result in the transfer of technology. In the absence of technology transfer, local production is at a disadvantage, as the development, testing, and registration of adequate formulations requires considerable technological skills and will usually take several years.

Even where there are sufficient reasons to reject the notion of creating new production capacity, manufacturing units that already exist may have a good reason to continue. Production units are frequently seen as strategic facilities with the capacity to promote economic development and to serve as training centers. In addition, even simple units may be capable of producing a limited range of finished products at reasonable cost. Where such units exist, for whatever reason, the authorities have a responsibility to ensure that production of medicines takes place according to GMP and that international quality standards are applied. Generics producers, especially exporters to developing countries (for example, India and China), have been playing a crucial role in ensuring price competition for key essential medicines, such as antiretrovirals. Ironically, within these two countries, domestic supply remains inadequate to meet current needs.

Financing

Problems in financing medicines supply exist at three levels. First, all poor countries experience a basic and chronic absolute lack of economic resources; second, resources that do exist are not always optimally allocated with sufficient

Although many developing countries promise universal access to essential healthcare, few can afford to deliver it

consideration for health needs within a country; and third, support from the international community has often been inadequate to supplement currently inadequate national resources.

On average, during 1997–99, the Least Developed Countries spent \$11 per capita on health, compared with \$93 per capita for lower middle-income countries and \$1,907 for high-income countries (CMH 2001). Although the minimum cost of a basic set of health services is still a subject of debate and there is significant variance among countries, there is increasing agreement that \$30 per capita represents the lower bound. Thus, although many developing countries promise universal access to essential healthcare to their citizens, few can afford to deliver it to their populations. As a result, in many poor countries today, much of the funding for healthcare is in the form of direct out-of-pocket expenditures by patients at the point of care. In wealthier countries, the public sector can raise money through general taxation or through social health insurance (employment taxes).

The current lack of international support in the health field to poor countries was already noted in the introduction to this report. The Working Group on Access to Essential Medicines fully endorses the findings of the Commission on Macroeconomics and Health (CMH 2001) and others (Troullier and others 2002) that donor assistance for health needs to rise dramatically in order to have an impact on the health of people living in developing countries. Others have noted that significant additional funding will be needed to target specific diseases. For example, the Institute of Medicine (United States) recently estimated that instituting new malaria treatment (artemisinin combination therapy) worldwide would require \$300–\$500 million each year (Coleman 2003).

In addition to inadequate aid flow, there is also a serious lack of coordination and transparency in donor assistance. In many cases, it is even difficult to estimate the total amount of funding going to support essential medicines. The lack of transparency and multitude of donors create significant transaction costs for poor countries, which must devote scarce staff to managing the morass of reporting requirements.

At the national level, it is evident that the money available for healthcare in general, and for medicines in particular, will often be strictly limited for many years to come. As noted above, in the short and medium terms, only significant additional donor assistance can ensure that countries can afford to offer basic health services, including essential medicines, to their populations. In addition, addressing access to medicines and other health issues will, in many cases, require a reallocation of priorities so that healthcare and medicine supplies are allowed to rank more highly on the scale of government expenditure, bearing in mind both the humanitarian and economic benefits that will flow from improved access to medicines. Detailing concrete actions to promote import (or production), distribution, quality testing, and promotion of sound prescribing within national poverty reduction strategies will be a key strategy

Evidence on the effectiveness of community insurance is mixed

for securing government and donor commitment to dramatically enhancing access to medicines.

Because of the low levels of government spending on healthcare, many patients are forced to pay out of pocket for health services and essential medicines. Out-of-pocket payments are the least equitable method of financing healthcare, as they are a disproportionately greater burden (as a percentage of income) for the poor than for the wealthy, and they come at a time when families are most vulnerable to usurious interest rates for medical loans. Furthermore, user fees, whether the full payment or a copayment for services, can act, for the poor, as a direct barrier to accessing needed health services. Even with extensive out-of-pocket spending for healthcare, the absolute spending is not enough because of the low per capita incomes of much of the populations in poor countries. For the poor, healthcare competes with other necessities of life, such as food and water, schooling, clothing, and shelter. Thus financing for essential health services in much of the developing world today can be described in two words: insufficient and unfair.

Some experts advocate community-based insurance as an important mechanism to offset the financial impact of healthcare on the poor. "Community insurance" is a term usually used to describe village- or district-level schemes to pool risk across the members by collecting monthly premiums from each family, which are then used to reimburse providers directly for costs incurred. In some cases, communities may employ the providers or have direct agreements with providers on fees for services—both of which can help to ensure a higher quality of services. Community insurance is preferable to user fees in that prepayments or premiums are often lower and more predictable, and they protect the family from catastrophic one-time payments when illness strikes. By transferring resources from the wealthier to the poorer and from the healthy to the sick, community insurance is also a more progressive payment mechanism than user fees.

Evidence on the overall effectiveness of community insurance as a financing mechanism is mixed. A 2002 review of 45 published and unpublished reports on community financing indicates that, while community insurance improves access to health services and reduces out-of-pocket health spending for poor rural communities, it often excludes the poorest of the poor, who simply cannot afford the premiums (Preker and others 2002). Another limitation of community insurance is its limited funding base and consequently limited coverage, which subscribers often prefer to direct to common illnesses requiring relatively cheaper types of care. As a result, community insurance usually does not cover rarer, high-cost, life-threatening events—precisely the ones that carry the highest risk of impoverishing a patient's family. Community financing schemes are also somewhat complex to administer, requiring a significant investment in management and oversight by communities. Community financing therefore is not, in the short term, a viable option for sustainable financing of primary healthcare in low-income countries (Ekman 2004).

User fees should be eliminated for essential health services and medicines

In view of this, the UN Millennium Project advocates for government-led financing of essential health services to ensure affordable access to a core set of services as guaranteed in the constitutions of many countries and in the Health for All initiative. User fees should be eliminated for essential health services and medicines. Financing is distinct from delivery of services, which can be done by NGOs and the private, for-profit sector, with appropriate regulation. Performance-based contracting is one promising tool for health service regulation. The UN Millennium Project also recognizes that current public health spending on healthcare is wholly inadequate in most low-income countries and meeting the Goals will require large inflows of donor assistance for health over the coming decade.

At the international level, there will often be a long-term need for donor support. Support itself should always (except in severe emergencies) have a development component and not be limited to supportive aid and the supply of consumables. Finally, as noted above, promises made about international aid need to be kept. Other factors that need to be addressed by the international community include punishing debt burdens and the imposition of social sector spending limits as a condition of loans from the international financial institutions. Support for debt relief in the Least Developed Countries has been gaining momentum in recent years, but responsive action on the part of the donor community has been inadequate. When seeking donor aid, states must quantify their needs for medicines and the extent of the shortfall in meeting those needs from national resources so that well documented requests for support can be developed.

Donors have traditionally been reluctant to fund recurrent costs, such as salaries. For their part, governments have been hesitant to rely on donors for such funding, given its lack of predictability. Yet providing adequate salaries for health workers is key to service delivery. A massive scale-up of services cannot happen without donor commitment to long-term sustainable funding for recurrent costs. The working group recommends that donors commit to funding salary and other recurrent costs for the poorest of developing countries over the short to medium terms.

Prices and affordability

Evidence of the role of price in obstructing access to medicines, in both the past and the present and in a wide range of countries, is abundant. Many institutions and authors have provided examples of gross discrepancies between the global prices of essential medicines and the ability of most ordinary individuals in developing countries to pay for them (Mossialos and Dukes 2001; WHO and HAI 2003). To cite a single example: in 2000, the costs of using didanosine for AIDS in the Côte d'Ivoire amounted to $3.48 per patient per day, yet the GNP per person was only $1.94 per day and the health services were able to make a contribution to the cost equivalent to only $0.03 per day (Mossialos and Dukes 2001).

The levers for pushing prices downward are generics competition, price negotiation, differential pricing, and effective procurement

Whether the price of a medicine is reasonable or unreasonable from the point of view of a manufacturer or supplier, if the patient or the health system cannot afford it, it will not be bought and used.

The problem is least severe for those medicines (such as generic products) that are off patent and can be obtained from multiple reputable sources. For many of these medicines, the international wholesale prices, based on generics supplies, are today only a fraction above manufacturing costs, though still sufficient to finance quality control, overhead, and distribution. These prices render the medicines accessible to all except the indigent, who will remain dependent on whatever support the social system can provide. Only if there is entirely inexpert procurement or if retail or other margins greatly inflate the price to the consumer do these medicines become unreasonably expensive.

Once a medicine has been taken into mass production, the costs of manufacturing it are generally extremely low. When the wholesale or retail price of a medicine represents a serious obstacle to its use, this price is primarily determined by factors other than the expense involved in making it. This allows substantial room for negotiation with the manufacturer. Certain medicines are reported to have a prohibitively high manufacturing cost, but this information is not publicly accessible. The example has been cited of natural insulin, which, even at the best generics prices, is still out of reach for poor people. Even here, however, it is not clear that, if synthetic human insulin were widely used, the cost would decrease sufficiently to be within reach of impoverished people.

Because newer medicines will be protected by patent from low-cost competition for at least 20 years, impoverished populations may (and generally will) be deprived of these medicines for that entire period. On the one hand, in the case of hypertension, rheumatism, and diabetes, this will not be a significant problem; medicines developed in the 1980s are still effective. On the other hand, patents will affect the affordability of medicines for prevalent diseases such as AIDS, MDR-TB, and multidrug-resistant malaria. Although research-based companies are actively engaged in donation or discounting and treatment programs (such as the Accelerated Access Initiative, which is treating 300,000 people in developing countries with discounted antiretrovirals), the sheer scale of the problem for AIDS alone precludes donations or multicompany discounting as viable strategies over time to counter the magnitude of need for treatment, however well intentioned or implemented.

The main levers for pushing medicines prices downward are generics competition, price negotiation (which can include the option of resorting to compulsory licensing), differential pricing offered by companies, and effective procurement practices, such as bulk or pooled procurement.

In almost all countries, a public sector and a private sector for medicines provision exist in parallel. This means that, in theory, an individual has two channels through which to obtain medicines, with different prices and charges in each sector. The public sector, often subject to serious funding constraints,

Equity pricing is based on the ethical notion that medicines should be less expensive in poor countries than in wealthy ones

will more readily pursue purchase of generics medicines to maximize purchasing power. The generally lower costs of these medicines are in turn passed on to patients, who receive the medicines either at no charge or at relatively low cost, sometimes in the form of a prescribing fee. The private sector supplies originator drugs at prices similar to those charged in Western countries, though retail prices may be quite variable within a country; most private channels also handle a certain number of generic medicines.

The inadequacy of public sector supplies in many developing countries means that much of the population will be obliged to use the private sector if they are to secure medicines at all, and it is here that serious problems can arise. Although most developing countries have some affluent citizens, they usually constitute only a small minority of the population. The bulk of users of the private sector will be those who have been either driven to use it or persuaded to use it and who can generally ill afford the prices that are charged.[6] A real risk is that high-cost originator medicines will be bought at the expense of other vital goods (such as food) or will be purchased in small quantities inadequate to serve their purpose, thereby perhaps doing more harm than good, for example, by inducing bacterial resistance (WHO 1998a).

Differential and equity pricing

The high prices charged in developing countries for essential medicines that are still under patent protection are a major barrier to accessing those products. Differential pricing, or tiered pricing, is widely practiced by the pharmaceutical industry as part of its marketing strategies. There are good examples of social responsibility leading to differential prices for essential medicines, including antiretrovirals. Yet it does not always result in lower prices for less affluent countries. For example, recent pricing surveys have shown that in many developing countries, such as Kenya, Morocco, and Peru, the prices of some orginator products are much higher than they are in the originator countries (WHO and HAI 2003). In some countries, taxes and tariffs further increase the price.

Equity pricing is a concept launched by WHO in the late 1990s. It is based on the ethical notion that developing countries should not be asked to pay for medicine development cost, marketing, and shareholder returns. This view that medicines should be less expensive in poor countries than in wealthy ones is widely accepted; the broadest shoulders can well carry the heaviest burdens. Equity pricing is a much wider concept than differential pricing and encompasses all the active policy and administrative measures a government or procurement organization can take to achieve differential pricing related to purchasing power. These measures include price information and transparency, bulk or pooled procurement, reduced taxes and margins, price negotiations, voluntary licensing agreements, and, as an ultimate measure, compulsory licensing. Equity pricing is the political choice and action; differential pricing

TRIPS created a comprehensive global patent regime for WTO member countries

may be one of the results. Equity pricing has been successfully practiced for more than 30 years for children's vaccines and reproductive health commodities, although care must be taken to ensure that multiple suppliers continue to participate to ensure continuity of supply.

In differential pricing, industry needs to provide these medicines at production cost ("no profit, no loss") to national health systems in low-income countries. In middle-income countries, differential pricing should be pursued, although the prices will not be at marginal cost. In both sets of countries, negotiating differential prices should not be burdensome to the purchaser. Differential pricing should not result in price referencing or the re-importation of the lower priced products in high-income countries. Administrative and regulatory measures are available to prevent this from happening. A good example in this regard is the market segmentation strategy for Novartis' Coartem®, with separate products for developed and developing countries, and separate presentations for the public and private sectors. However, it must be borne in mind that that it is not a simple process to use administrative and regulatory measures to offset the negative consequences of differential pricing. There is a significant cost burden associated with maintaining separate brands, presentations, and packaging and other issues, in terms of both direct costs and human resources.

The need for timely pricing information

The WHO, in collaboration with HAI, has developed a price survey methodology that is an important step forward in understanding what patients actually have to pay. It is a useful tool for collecting and comparing prices, especially at country level, helping policymakers to address high and variable pricing of essential medicines. The WHO also collaborates with MSH and MSF to produce information on prices offered by various suppliers.

Intellectual property protection

The world's network of patent systems was, until late in the twentieth century, far from absolute or homogenous. Some nations had no patent systems at all, while in others patents were not consistently enforced. This enabled low-cost manufacturers in some countries to produce new medicines legally. They were sometimes manufactured purely for internal sale, but were also commonly exported. Following the establishment of the World Trade Organization (WTO) in 1995, its agreement on Trade-Related Aspects of Intellectual Property Rights (TRIPS) created a comprehensive global patent regime for WTO member countries (WTO 1994; Drahos and Braithwaite 2002). Implementation of TRIPS began in 1995 for developed countries (which were also countries most likely already to have extensive patent protection). Less-developed countries were put on a rolling schedule of when they had to bring their national legislation into compliance. Least Developed Countries

The schedule for TRIPS implementation was probably not well grounded in economic and development realities for poorer countries

originally had until 2006; through subsequent negotiations, they now have until 2016 to offer patent protection for pharmaceuticals. TRIPS recognizes that some flexibilities would be required to allow governments the means to respond to domestic requirements, including for public health considerations.

As the UN Millennium Project Task Force on Trade has pointed out, the schedule for TRIPS implementation was probably not well grounded in economic and development realities for poorer countries:

> There is perhaps more agreement about the extent to which the TRIPS Agreement provides sufficient flexibility for developing countries. As a basic matter, there is wide agreement that the time and resources required to implement the Agreement were greatly underestimated and that implementation has (and will, if nothing is done) put a considerable strain on many developing countries. Assistance from developed countries, and the additional implementation periods permitted developing countries, have not been commensurate with the size of the task. (UN Millennium Project 2005a, ch. 10).

Another problem the Task Force on Trade highlights is that of flexibility required to account for different levels of development and national priorities for development:

> Additionally, in some cases, the substance of the Agreement provides insufficient flexibility, imposing a "one size fits all" model of [intellectual property rights] protection on countries at widely differing levels of development and requiring protection of the full range of [intellectual property rights], despite varying interests and priorities. In other cases, the problem may be not so much that the Agreement has no in-built flexibility. Rather, it is that some WTO members are not permitting others to take advantage of the existing flexibility. For instance, while the agreement provides for differing implementation periods, countries acceding to the WTO may not even have access to these normal flexibilities. Additionally, certain WTO members—the US on drugs, the EU on [geographical indicators]—are trying to impose strict (and thus unacceptable for a vast majority of the rest of the world) limits on the existing TRIPS flexibility. (UN Millennium Project 2005a, ch. 10)

It has been recognized for many years—and is today virtually unquestioned—that if enterprise and innovation are to be encouraged, the innovators must be in a position to exploit their discoveries so as to reap their due reward and finance future innovative work. To this end, the issuing of a patent on the discovery of a new medicine, generally for a period of 20 years, provides innovators with protection from competitors. In general, the patent system works well because the inventor will have every reason to make a discovery widely available to the community through production and licensing and to publish

Most countries allow the government to use patented inventions for public purposes with fewer bureaucratic obstacles than apply to the private sector

patent information free from the threat of competition; theoretically it will therefore benefit both the inventor and society.

Although the patent system has developed primarily in Western industrialized countries, it can also provide an advantage to emerging market economies and some developing countries in advancing R&D in various fields. Such considerations have led to an almost universal acceptance of the patent principle, whatever problems it may pose in particular situations.[7] One of these problematic situations involve the issues surrounding inadequate or absent access to medicines.

The TRIPS agreement does embody a number of provisions for exceptions to be made to its rules, and these can be relevant to the issue of medicines pricing. These inherent flexibilities were endorsed in the November 2001 Doha Declaration on TRIPS and Public Health, approved by all 146 WTO member governments and endorsed by the research-based and generics industries (WTO 2001). There is every reason, where medicines are concerned, to exploit these exceptions—targeted to address the constraints of developing countries—to the full (Love 2000). It should be noted, however, that the interpretation of these exceptions in the field of medicines has been a matter of dispute since the Agreement was concluded, and that the extent to which these exceptions will be of value is still in doubt.

A potentially broad exception provided for by TRIPS itself is to be found in Article 30, which states:

> Members may provide limited exceptions to the exclusive right conferred by a patent, provided that such exceptions do not unreasonably prejudice the legitimate interests of the patent of the patent owner, taking account of the legitimate interests of third parties. (WTO 1994)

An example of such an exception is the Bolar clause, which allows for fast introduction of a generic after the patent term by permitting technical preparation for registration of the same medicine from an alternative source before the patent has expired.

A further vital tool for the governments of developing countries to use in dealing with obstacles presented by patents is Article 31 of the TRIPS agreement, which sets out the procedures for compulsory licensing and government use of a patent. A compulsory license is an authorization by the government for itself or a third party to use that patent without the permission of the patent holder. Most or all countries—developed and developing—allow the government to make use of patented inventions for public purposes with fewer bureaucratic obstacles than apply to the private sector. A compulsory license authorizing the government to use the patent for its own purposes is also referred to as a government use authorization.[8] In practice, compulsory licenses are not normally used. Most often, it is the presence of the ability in national legislation and the threat to invoke it that results in desired price concessions

The Fourth WTO Ministerial Conference in 2001 affirmed the right of national governments to take measures to protect public health

from suppliers. There remains an obligation to pay the patent holder "adequate remuneration in the circumstances of compulsory licensing or government use, taking into account the economic value of the authorization" (WTO 1994).

A limitation on the use of Article 31 is the requirement that governments proposing to exploit this exception seek in advance the agreement of the patent holder to use the invention on reasonable terms (such as to seek a voluntary license). A compulsory licence can be issued only if such agreement cannot be obtained "within a reasonable period of time" (clause b). However, this requirement may be waived by a member in the case of "a national emergency or other circumstances of extreme urgency or in cases of noncommercial use" (WTO 1994). Countries are free to determine what constitutes a national emergency and do not need to follow any official procedures.[9] It is not, for example, obligatory to officially declare that a state of emergency exists.

Countries are also free to define what constitutes public noncommercial use. This can, for example, be defined as covering procurement or production of healthcare products for use in the public sector. In practice, this means that a procurement authority in a country can start the purchase of generic versions of needed medicines without prior negotiations with the patent holder. The patent holder will be informed of the decision to make government use of the patent and the government will have to offer to the patent holder adequate compensation, the level of which is determined by the government itself. Article 31 further rules that such use "shall be authorized predominantly for the supply of the domestic market of the Member authorizing such use" (WTO 1994).

The Fourth WTO Ministerial Conference, held in 2001 in Doha, Qatar, adopted the Declaration on TRIPS and Public Health, which affirmed the right of national governments to take measures to protect public health and appeared to legitimize the broad use of these flexibilities, including compulsory licensing (Abbott 2002) where medicines were concerned (WTO Council for Trade-Related Aspects of Intellectual Property Rights 2001). With opposition from some industrialized countries, the declaration is weaker than originally proposed and is not legally binding. Nevertheless, from a legal perspective, it is an important document that will have to be taken into account by any WTO panel dealing with this issue in practice.

The Doha Declaration has also created new rights, for example, the right of Least Developed Countries to exclude pharmaceutical products from patenting until 2016. This right is legally binding and cannot be challenged (figure 2.1). It has been criticized by some defending the point of view of the research-based industry (Gillespie-White 2001), but others have seen it as a major step to protect the interests of developing countries (Correa 2002). The generics industry has welcomed this protection, while pointing to the problems that remain.

The latest developments are those resulting from a decision of the WTO General Council on August 30, 2003. A key issue that remained unresolved at Doha was how to ensure production for export to a country that has issued a

Figure 2.1
August 30 waiver scenario for a Least Developed Country

Source: World Bank 2004a.

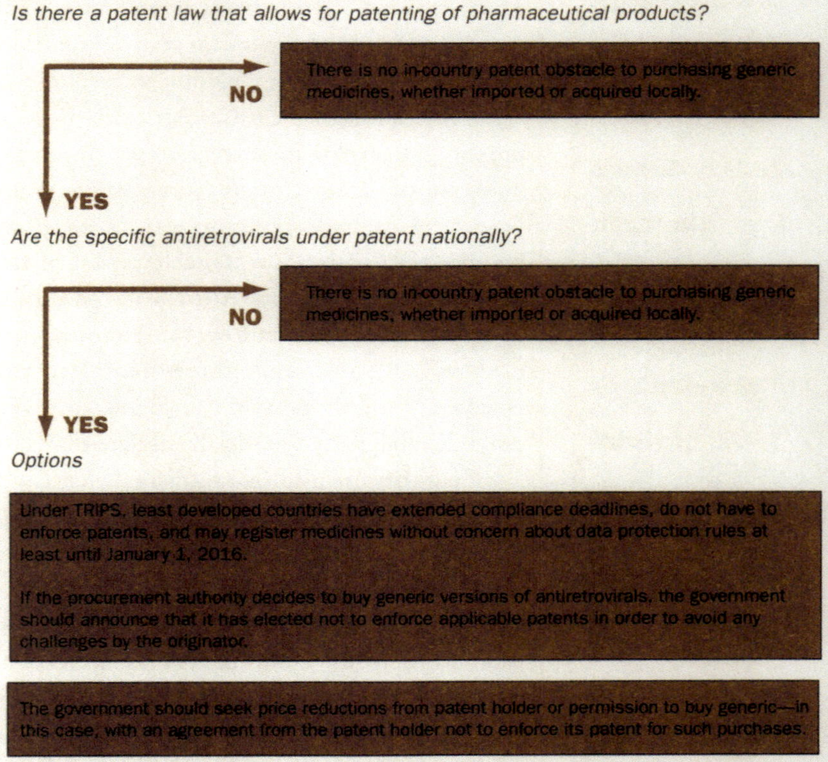

compulsory license but that does not have adequate manufacturing capacity. Since Article 31(f) of TRIPS limits compulsory licensing to uses that are predominantly for the supply of the domestic market, WTO members agreed that further action was necessary to ensure that countries without production capacity can use compulsory licensing provisions to the same extent that countries with production capacity can use them (WTO Council for Trade-Related Aspects of Intellectual Property Rights 2001). The Doha Declaration on TRIPS and Public Health acknowledges the problem in paragraph 6, and on August 30, 2003, the WTO adopted a decision on a waiver to the 31(f) requirement. Concurrent with this agreement was the requirement that exporting producing countries also have to issue a license (which will require amendments to their existing domestic legislation). Overall, the system appears to be very cumbersome, and it may be beyond the administrative capacities of many developing countries to use effectively (figure 2.2). More will be known about the usefulness of this agreement after 2005.

Since the adoption of the Doha declaration, the 32 Least Developed Country members of the WTO have been able to benefit from exceptions with regard to pharmaceutical product patents. Paragraph 7 of the Doha Declaration provides a special extension of the TRIPS transitional period for pharmaceutical products. Least Developed Countries do not have to "implement or

Figure 2.2
August 30 waiver scenario for a developing country

Source: World Bank 2004a.

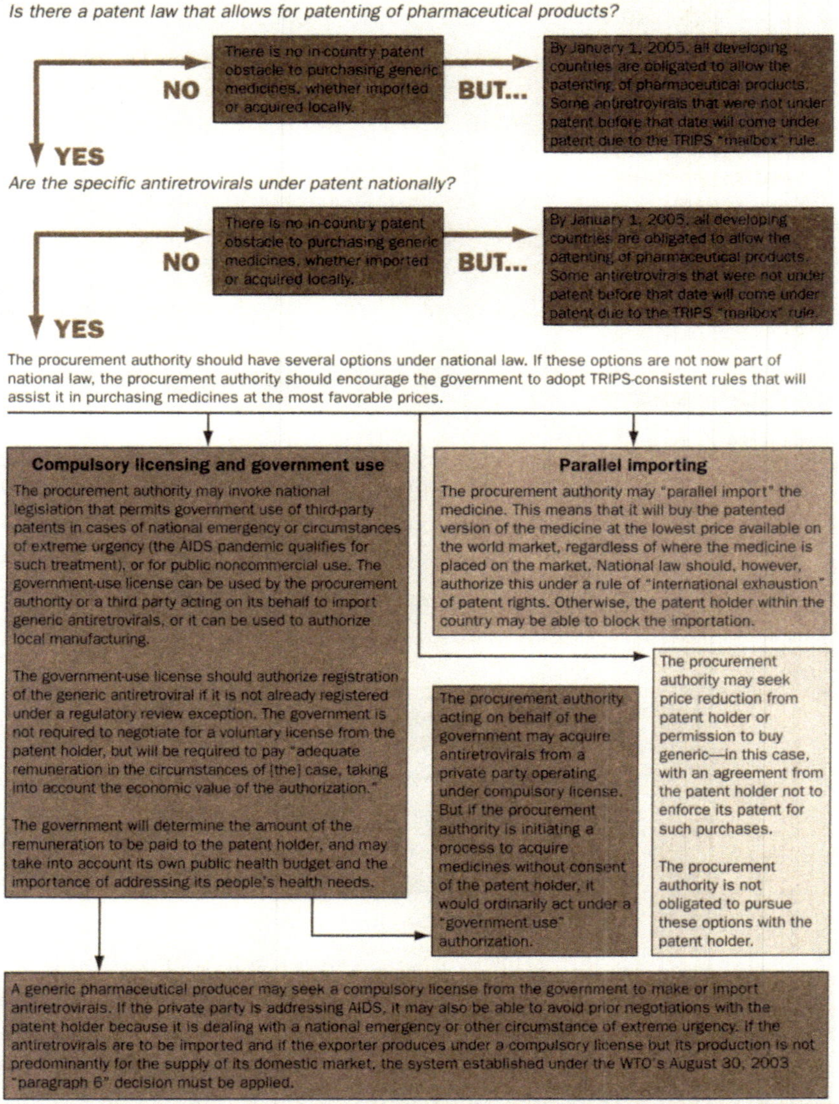

apply Sections 5 and 7 of Part II of the TRIPS agreement or to enforce rights provided for under these Sections until 1 January 2016" (WTO 2001). This means that Least Developed Countries do not have to grant product patents for medicines, provide protection of undisclosed test data, or enforce patents that have already been granted until at least 2016.

The UN Millennium Project Task Force on Trade has concluded that the impact of TRIPS on access to essential medicines will probably have a negative effect over time on developing countries:

> From an economic perspective, the first-best approach to intellectual property in the drug industry would be to subsidize research and development and to grant no patent rights (or to grant patent rights

It seems clear that free trade agreements will negatively affect access to new patented medicines and vaccines in developing countries

and subsidize production up to the point where the patent-holder maximizes profits by setting price equal to marginal cost). However, this approach seems out of reach for a long time.

In fact, the international extension of intellectual property rights, as it applies to drugs, may progressively become (as the effect of TRIPS-induced patent protection will be felt mainly in the future) welfare-reducing from a world perspective and particularly from a developing country point of view. This is because most developing countries have virtually no ability to contribute meaningfully to the costs of developing major drugs, and there is little worldwide gain in terms of new product development funded by developing country purchases.[10] By contrast, the cost of drug protection to developing countries may increase because the monopolies created by the extension of patent protection may progressively cut many developing countries off from essential medicines. In sum, no innovation gain may ultimately compensate the monopoly-related loss brought about by extending patent protection to the developing countries. (UN Millennium Project 2005a, ch. 10)

Regional trade agreements, TRIPS-Plus, and access to medicines

According to the UN Millennium Project Task Force on Trade, virtually all WTO members are party to one or more free trade agreements. The WTO estimates that, by the end of 2005 the total number of free trade agreements in force might well approach 300. The proliferation of free trade agreements is compounding concerns about the impact of trade agreements on access to medicines (UN Millennium Project 2005a). Box 2.1 summarizes the various ways in which free trade agreements are imposing TRIPS-Plus conditions that affect access to medicines.

In reviewing the evidence on the impact of trade agreements on access to medicines, it is a clear likelihood that these agreements will negatively affect access to new patented medicines and vaccines in developing countries. Although some steps have been taken to address undue limits on the existing TRIPS agreement (such as the waiver for Article 31[f]), it is likely that more will need to be done as more becomes known about the effects of implementation after 2005. At the least, developing countries will need more technical assistance from a range of expert sources to cope with the technicalities of trade agreements at the country level. While debate continues in some quarters about the impact of patents and trade agreements on access to medicines, it is safe to conclude that the proof will not be long in coming, since 2005 is a milestone year for the implementation of TRIPS in key countries and for the launching of some free trade agreements that impose further restrictions on how public health needs can be addressed. It will be important for the WHO,

Box 2.1
Examples of TRIPS-Plus provisions in free trade agreements

Sources: Oxfam 2003, 2004; Vivas-Eugui 2003; Drahos 2004.

Extension of patent protection beyond the 20 years required under TRIPS. Patent terms should be extended to compensate patent holders for any unreasonable delays in granting the patent or unreasonable curtailment of the patent term as a result of the marketing approval process. There are not such requirements under TRIPS and thus the effective period of protection under TRIPS is usually less than 20 years.

Limits on parallel imports. The patent holder is permitted to restrict the possibility of parallel imports in the market. TRIPS is silent on parallel importation.

Test data protection. Test data of patent owners must be protected for at least 5 years for pharmaceutical products (10 years for agricultural chemicals) from the date of approval of the patent, delaying the marketing approval of generic drugs. Should this requirement continue to apply even where a compulsory license has been issued, it would effectively prevent the use of such licenses as the delay and costs would be too great. TRIPS requires protection of such data only against "unfair commercial use."

Compulsory licensing. The grounds on which compulsory licenses can be issued are more restrictive than they are in TRIPS, and requirements for compensation to the right holder may be higher than required by TRIPS.

Marketing approval and the life of the patent. Requirements exist to disclose the request for marketing approval and identity of the applicant to the patent owner; patent holders are alleged to use frivolous lawsuits to unnecessarily delay marketing approval for generics. TRIPS permits generic producers to seek regulatory approval during the life of the patent with no conditions.

perhaps in cooperation with other agencies, such as the WTO, to monitor TRIPS implementation and free trade agreements as they pertain to medicines in coming years and to report, initially by the end of 2007, findings to date and recommendations.

Data exclusivity and evergreening

Although the degree of protection conferred by patents, generally for a period of 20 years, is clear, one must also take into account a number of techniques that originators have used or sought to use to defend or extend their rights.

A widely used way to obtain additional patent rights beyond the patent term for the original compound patent is to enhance the original product in some way so that a new patent may be granted on the new invention (NIHCM 2000; MSF 2003a). Sometimes the modification itself constitutes a significant innovation of importance in health terms; other times it may provide only marginal benefit in terms of usefulness, efficacy, or safety, and thus the modified product is unlikely to replace the original except to the extent that prescribers move to it.

There is also an ethical aspect of extended data exclusivity. Many patients voluntarily participate in clinical trials and accept the inherent risks associated with that participation, with the understanding that they contribute to the benefit of future patients and in the interest of the advancement of medical

Price transparency is an essential tool for designing public policy, promoting competition, and keeping prices in hand

science. However, they could be more hesitant to participate if they knew that the outcome of the trial would not become part of the public good and would only benefit one commercial company to the exclusion of all others. An increasing number of scientists and consumers are arguing that the data on which the efficacy and safety of medicines are being assessed by regulatory agencies should be open to public scrutiny. Some research-based companies are beginning to act to increase the transparency of clinical trials data, for example by participating in online registries.

Procurement

Countries with scarce resources use a variety of ways to procure medicines. Whenever a developing country does find ways to reduce the cost of the medicines it procures, it can pass the benefits on to its population through reduced prices or fees and broader access.[11]

It is clear that governments and their agencies have a leading role to play in negotiating or agreeing to the prices at which medicines can be acquired. In developing countries, central medical stores or similar bodies commonly play that role.

In the public sector, procurement of medicines may be either insufficient because of lack of resources; inefficient because of lack of information, expertise, or negotiating power; or simply inappropriate to the country's highest-priority health needs. In the private sector, some medicines are also inaccessible for a given country because there is no commercial channel able or willing to import them. In such cases there may be no alternative to the public sector filling the gap insofar as it is capable of doing so.

Lack of information on prices and sources is a problem that can be solved relatively simply. For many essential medicines, various impartial bodies have issued compilations of the prices at which medicines and diagnostic supplies are available on the world market; such guides provide valuable support to procurement bodies.[12] Available evidence shows that price transparency is an essential tool for designing public policy, promoting competition, and keeping prices in hand. Special offers are available for some originator drugs. MSF has provided an overview of discounts, donations, and other offers available from manufacturers and the conditions attached to them (MSF 2003e); here too industry has criticized the data used, but at the country level, the positive impact of such documentation on procurement is evident.

Lack of information on the population's needs from district to national levels will require focused operational research and the use of available epidemiological statistics to overcome. Drug utilization studies providing data to guide procurement are needed everywhere and in a form that is relatively simple to carry out using readily available data (Dukes 1993). Technical support in this field can be provided by the various regional drug utilization research groups sponsored by WHO.

Box 2.2

Some conditions for the success of regional pooled procurement schemes for medicines

Source: Adapted from SEAM 2003.

- Homogeneity of member states: size, range of needs, economic development, culture, political tradition, language.
- Harmonized national requirements: drug regulation, taxes, import duties.
- Financial stability: stable currencies, countries able to pay for pooled services and for supplies received.
- A common approach to quality: agreed quality standards, agreed procedure for control of suppliers and batches.
- Reasonably accurate prediction of needs.
- Competent and stable central staff.
- Reliable data on the patent situation of medicines.
- Loyalty of member states: national procurement agencies must not compete with the pool.
- Monitoring performance at pool and national level.

Lack of expertise in international procurement and in negotiating prices is a difficulty experienced in the public sector in many smaller countries. Donors have often provided support by training procurement officers. However, well trained and experienced individuals in this field are often lost to the private sector, where salaries are much higher. A novel approach to public procurement, which has been adopted in various parts of the world in recent years, is pooled procurement, an arrangement by which a number of countries jointly entrust their drug purchasing to a single body, generally working at the regional level.

The advantages of pooled procurement systems include:

- Access to experienced negotiating expertise and market knowledge.
- The ability to purchase medicines from the supplier on a larger scale, frequently resulting in significant bulk discounts.[13]
- Substantially widened access, since major suppliers that would not ordinarily tender for very small national markets are willing to tender to the larger regional procurement bodies.
- A provision of central financial guarantees to suppliers that will apply should any member state default on payment.

Essentially, a regional pool takes over the task of providing all medicines for its member states, either across the board or within a defined therapeutic area. The pool remains dependent on the individual member states for forecasts of need, if necessary assisting member states to undertake these forecasts, so that ordering is adjusted to real requirements. Some schemes go further: the regional procurement body operating in the Eastern Caribbean provides considerable support to its member countries in such matters as encouraging good prescribing by well constructed formularies, hence promoting the most efficient use of resources. (See box 2.2 for a list of conditions for success of regional pooled procurement strategies.)

The following excerpt is from the 134th Session of the PAHO Executive Committee, *Report on the 38th Session of the Subcommittee on Planning and Programming,* June 2004:

A range of specialized global funds and agencies have entered the field of global procurement

> In examining options for pooled procurement, it would be important to review the experience with the PAHO Revolving Fund for Vaccine Procurement, applying lessons from that to the development of other procurement mechanisms. The Revolving Fund had initially, in 1979, been capitalized at $1 million, with 19 countries participating, for the purchase of five vaccines. The corresponding figures now were $24 million, 35 countries, and 12 vaccines, and the Fund had contributed significantly to the achievement of priority objectives in immunization in the Americas by supporting countries in commodity procurement, supply, and use. Another option for pooled procurement was the PAHO Strategic Fund, which had been established to help countries in the procurement of HIV/AIDS, TB, and malaria products. Although 11 countries had signed participation agreements with the Fund, it had been used in only a few countries to date. The Fund could become an effective instrument for ensuring continuous supply of public health products and building capacity in supply management, but in order for that to happen it would be necessary to reaffirm political commitment, redirect the technical cooperation package supporting the Fund, review the administrative procedures governing operation, and develop lines of communication with countries. (PAHO 2004, p. 20)

A different approach is involved in the global procurement of medicines. These initiatives have been undertaken at various times in order to meet a major worldwide need for a specific type of product. The best known of these involved procuring oral contraceptives and vaccines for a range of countries, with UNICEF and the Rockefeller Foundation taking the lead, and securing prices that represented only a small fraction of the supplier's usual market price. UNICEF has for many years maintained a nonprofit global supply system for drugs, vaccines, and other essential supplies, with its own procurement agency and its own warehousing and delivery program. UNICEF has claimed that, although it supplies some 40 percent of the vaccine market in unit terms, it covers only 5 percent of the market in financial terms (Jarrett 2003), which would indicate that its vaccines are procured at only a small fraction of the prevailing industrialized country market prices. It also indicates that they are supplying poor countries rather than pricing and supplying to markets in wealthy countries. More recently a range of specialized global funds and agencies have entered the field of global procurement, for example, for AIDS. Established in 2001 by the Stop TB Partnership, the Global TB Drug Facility provides grants in kind to some countries and technical support for procurement in others.

It has been argued that a very large pooled procurement scheme might actually exert excessive downward pressure on prices, thus rendering the market so unattractive that some suppliers would withdraw (Jarrett 2003). This

Technical pharmaceutical regulatory systems are essential means of ensuring that medicines attain the necessary standards of efficacy, quality, and acceptable safety

risk probably arises only in the case of a very large and powerful joint procurement agency with global outreach, but it should be borne in mind.

Regulation

The regulation of medicines exists to protect public health. In theory, regulation should not create a barrier against access to bona fide medicines, but in some circumstances it can do so. Bureaucratic delays can occur or excessive demands may be imposed, delaying distribution or increasing systemic costs. Regulatory regimes and measures are challenged on such grounds by both research-based and generics companies, and it can be helpful to consider some of the problems that arise.

Basic standards

Technical pharmaceutical regulatory systems are essential means of ensuring that medicines entering the market attain the necessary standards of efficacy, quality, and acceptable safety, and that the information provided with them is sufficient and reliable. In certain areas of the world (most notably the European Union, but also in parts of Africa and members of the Association of Southeast Asian Nations), collaboration between agencies has been established and has progressed toward regional regulation. Particularly where agencies are small and understaffed, as is the case in much of the developing world, it is likely that this development will reduce unnecessary delays.

National regulatory agencies need to be strengthened. They may wish to consider developing fast-track procedures for medicines for priority diseases in a given country. WHO prequalification offers national agencies a reliable source of information about priority medicines for AIDS, TB, and malaria.

Equivalent versions of medicines

Because of the commercial value to research-based companies of the medicines that they have researched, developed, and marketed, understandably bitter disputes have arisen around the licensing of generic versions from other firms. These medicines, developed by copying the active ingredient, allow firms to benefit from all the creative effort of the originator. Insofar as the original medicine is patented, exact copying will not be possible so long as the patent remains valid. It is not, however, the task of a regulatory agency to determine, when considering an application for regulatory approval, whether the medicine in question complies with patent law or not.

In general, the generic version of a product should be bioequivalent to the original. Not all agencies have a common policy regarding the evidence that will be required to demonstrate bioequivalence, and the level of proof required will depend to some extent on the nature and form of the drug.[14] The essential principle is that bioequivalency can be said to exist only when the product in question will, beyond reasonable doubt, have precisely the same kinetics and

A national government must ensure that there are effective means of supplying and distributing medicines to the entire population

effects in an individual in the same dose as the original. This will usually require testing on a living subject. It would be desirable to arrive at clear and universally agreed criteria for determining when this bioequivalence can be said to exist.

Similar medicines

Regulatory agencies are regularly confronted with medicines based on new active substances that are so close in chemical structure to those already known that a hypothesis arises that they will have the same effect. As a rule, this can be no more than a hypothesis, since even slight differences in chemical structure can result in major differences in pharmacological activity and effects. While such a drug should be subject to full regulatory requirements, it is generally accepted that many agencies have, to some extent, applied less stringent criteria in the review process.

Taxation

Import duties or taxes, imposed by another government department, may lay an excessive burden on medical supplies. Delays at customs can also mean that medicines lie unused for long periods in port facilities, sometimes actually expiring during this time or being subjected to suboptimal storage conditions that can cause degradation in their quality because of exposure to excessive heat, cold, humidity, or light.

A value-added tax (VAT) is a revenue-raising instrument that can exist at several different levels of the system and may be applied to different classes of products, including, in many countries, essential medicines. A 2003 European Commission study found that VAT rates imposed on medicines averaged more than 12 percent (Irvine 2004). The combination of duties and taxes can significantly increase the retail price of medicines. While the global average increase is 18 percent, for many low-income countries the increase is higher; for example: India 55 percent, Sierra Leone 40 percent, Nigeria 34 percent, and Bolivia 32 percent.

Distribution

Medicines entering a country may be reasonably accessible only in urban centers because of a lack of a countrywide distribution system. Private distribution systems (through wholesalers and pharmacies) operate only in the urban areas in many developing countries. Public systems, set up to provide national coverage, often experience chronic or incidental problems: management and ordering routines may be poor; transport networks may be irregular or incomplete (especially in areas with poor communications); and losses may occur due to poor storage, theft, or corrupt practices at one level or another. Lack of qualified and dependable staff exacerbates all of these problems.

A realistic interpretation of the duties of a national government in this field is that it must ensure that there are effective means of supplying and

Taxes and tariffs on essential medicines should be eliminated; they negatively affect both affordability and competition

distributing medicines to the entire population. There is no conclusive evidence to support the superiority of public over private distribution systems or vice versa. In many countries the two co-exist; sometimes one has developed because of defects in the other. The entire balance between private and public operation must be subject to ongoing review, and the system of control adjusted as necessary to counter shortcomings identified in the system. Where a public supply system is proving unsatisfactory, the possibility of transferring operations to the private sector or subcontracting certain tasks (such as procurement, transport, or administration) may need to be considered.

A special place can often be accorded to faith-based NGOs with their own distribution systems. Some of these, such as Joint Medical Stores in Uganda, provide exemplary models for efficient, low-cost operation. Kenya is one of several countries where the most successful medicines supply system (that is, one providing medicines to the mission health sector) has been managed by an NGO rather than a government agency (WHO 1997; Kawasaki and Patten 2002). The country's Mission for Essential Drugs and Supplies was set up in 1986 by the Catholic Secretariat and the Christian Health Association to supply medicines to church-managed health units. Financial support has been provided by various bilateral donor agencies, but by 2002 the system was found to be financially self-sustaining. Supplies are procured in bulk from local agents and local producers and the system maintains its own facilities for storage, distribution, and quality control. Monitoring for efficiency is intensive, with operating expenses averaging only some 10 percent of total costs. There have been extensive training programs for health facility staff, though these cover only part of the training needed.

Competition

The means to promote and regulate competition normally operates at the national and local levels. On the national level, it will be heavily influenced, and to some degree regulated, by features such as national licensing policies, importation regulation, and other public policy tools. The degree to which this does or does not regulate competition will also vary (at times significantly) from country to country. Taxes and tariffs on essential medicines should be eliminated; they negatively affect both affordability and competition.

On the local level, manufacturers, importers, and distributors may or may not be constrained in setting the terms of commerce (Huttin 1994). Within any country, the promotion of competition is a potent tool to ensure that prices fall to a fair level. It should include competition between various therapeutic approaches and between orginator and generic medicines producers. The prices charged for marketed medicines at all levels—imported, wholesale, and retail—must be published and constantly monitored so that excessive charges can be detected and eliminated; this is already the practice in many industrialized countries and it should be applied everywhere.

Establishing and maintaining standards of good quality are critical for reliable access to medicines

Supply and distribution in developing countries could, theoretically, be addressed by the private sector. However, in most low-income countries, the private sector is not robust enough to attract this type of investment. The public sector, in the form of central medical stores and similar bodies, arose precisely because the market was not responding to national needs in terms of imports, distribution, and price. It is worrying that the United States appears to be contracting private international firms to deliver medicines supplied through its PEPFAR program. Such approaches, while technically "private," have a chilling effect on the development of local, private, competitively based distribution mechanisms.

In developing countries, it is not unusual to find a profusion of private retail pharmacies, often clustered within a single section of a major urban center, that have developed in response to the demand for expensive originator medicines by a wealthy minority. Beyond this urban setting, drugstores (which are often poorly stocked) may or may not exist. Despite rapid urbanization trends, the majority of the world's poorest populations continue to live in rural, often very isolated, regions. Competition, for the most part, is nonexistent in these areas because there are no medicines to be obtained by any means. Competition is severely constrained because of a lack of demand (because people cannot afford to pay) and high costs of supplying physically isolated locations.

Patent protection can also be viewed as a barrier to competition (this assumes serious dimensions when a medicine is virtually unique and irreplaceable) during the period of patent protection. Unless or until a suitable alternative medicine enters the market, there will be no genuine competition capable of reducing prices. However, experience has shown that in some cases, good-faith negotiations with the patent holder have led to expanded access to needed patented medicines on acceptable preferential terms.

Quality

Once a product has been developed and approved, quality manufacturing is required to bring it to consumers. It is evident that if a medicinal product is of poor quality it cannot realistically be regarded as accessible. The failure to adhere to adequate quality standards may result in immediate or long-term injury to human health. The importance of establishing attainable standards of good quality (quality assurance) and of ensuring that these are maintained during production (quality control) becomes a critical factor in reliable access to medicines.

A problem from the global point of view is that the experiences of agencies in this matter are generally not published, and it is difficult to determine how watertight the methods are and where and to what extent problems are encountered (Kaplan and others 2003a, 2003b). The same applies to some excellent studies known to have been conducted by bilateral aid agencies; it would be helpful if these were to be made public. Certainly, however, there is abundant

Only a minority of medicine manufacturing facilities are as yet producing medicines to GMP standards

evidence that widespread quality problems persist even where some life-saving medicines are concerned, such as antimalarials in Africa (Maponga and Ondari 2003).

Quality standards

There can also be some difference of view on the quality standards to be applied. Standards for older medicines are usually to be found in national pharmacopoeias. During the last 20 years, the concept of GMP standards (supported by WHO) has come into use. There are also, however, stricter standards, such as those propagated primarily through the International Conference on Harmonization (ICH).

Representatives of drug regulatory agencies and the pharmaceutical industry from the European Union, the United States, and Japan first developed the ICH in 1990. First proposed as a regional initiative to eliminate duplication of regulatory efforts and achieve quicker access to new pharmaceuticals, global expansion of the initiative has been planned since 1997. During a meeting of the Neglected Diseases Group in Malaysia in February 2004, concerns about the impact of ICH regulation on access to medicines in developing countries was discussed. In addition to the use of the ICH as a global standard without any clear international mandate or any international harmonization, other concerns echoed those expressed during a meeting in Geneva in 2003. One of the main concerns was that it would increase the costs of raw materials and generic medicines without any quantified increase in quality and therefore without clear public health benefits (Bannenberg 2004).[15] The standards are justified in certain situations, but they are often complex and disproportionately costly and have been criticized by independent experts as unnecessarily strict.[16]

Of the many thousands of medicine manufacturing facilities throughout the world, only a minority are as yet producing medicines to GMP standards. Many more must be induced to do so. Achieving GMP is often an incremental process that, given the economic and logistical constraints inherent in many developing countries, will take longer for some manufacturers to fully implement than others. This will be achieved only if procurement agencies are firm in insisting on these standards as a condition of purchase. Where these standards have not yet been attained, the decision will have to be made from case to case about whether a particular product or supplier offering lesser standards can, as a temporary measure, be regarded as acceptable. As noted above, much will depend on the nature of the medicine concerned; no flexibility can be allowed for medicines with a narrow therapeutic margin, but some flexibility may be tolerated for medicines with a broad safety margin.

Generic medicines

The misperception that generic medicines are inherently of lower quality than originator products still exists. In fact, generics must demonstrate the same

Quality controls exercised by manufacturers are complemented by those carried out by public authorities

quality standards as the originator version in order to be registered. Ongoing quality control criteria remain the same for either type of product. It is a fundamental fact that approved generic medicines are identical in their effects and usefulness to the corresponding products of originator companies. An approved, chemically derived generic pharmaceutical is thus medically interchangeable with the originator product and with all other generic clones of the original patented medicine. For many products, proof of this interchangeability will require a bioequivalence study. For some products and in certain countries, this requirement may be bypassed where other guarantees of identity are considered sufficient. In this respect and in other regulatory matters it must, however, be acknowledged that some resource-short nations do not succeed in enforcing such standards. Where that is the case it is vital that the national authorities "stand tall" politically and allow, for as long as necessary, an international agency to assist them by coordinating and simplifying the regulatory and approval process, and where necessary assume some technical functions that they cannot perform themselves (for example, the inspection of foreign manufacturing plants and products, as is done by the WHO for antiretrovirals for AIDS, antimalarials, anti-TB medicines, and other essential medicines) (Haddad 2004).

Official quality controls

Quality controls exercised by manufacturers are complemented by those carried out by public authorities. Many countries maintain quality control laboratories of different sizes and levels of competence, both in order to examine quality standards at the time of registration and procurement, and to check suspect samples from the field.[17] Any importing country should, in principle, be capable of protecting its population from generic products from unreliable sources by relying variously on its own regulatory approval systems and inspectorate and on international systems for cross-border inspection and prequalification of firms. The costs of maintaining such a system are not negligible, but they are likely to be modest when compared with the savings that can be achieved in terms of reduced waste.

Prequalification of suppliers and products

For many years, some national and other procurement agencies have limited their purchasing to prequalified suppliers, that is, firms or individual products that they have investigated in advance and found to be of sufficient standard. However, the fact that a supplier has been prequalified provides no absolute guarantee that the products that he or she supplies will in all cases meet the requisite standard; ongoing quality control at the time of supply remains essential. An ambitious prequalification scheme is currently operated within WHO to serve a number of agencies purchasing drugs for AIDS, TB, and malaria (Quality Assurance and Safety of Medicines 2003). Key steps in the process include assessing product dossiers for safety, quality, and efficacy and assessing

Bona fide quality generic medicines have sometimes been erroneously regarded as counterfeit or substandard items

manufacturers for compliance. Here, however, the same reservations apply: the WHO has had to remove certain prequalified products from its list when it has been found that they do not fully meet the required standards (some have been relisted after submitting more data on bioequivalence). Prequalification is thus an approach that can save time and can generally be reliable as a way to select suppliers and products for a particular order, but it should be part of a larger quality assurance strategy.

Substandard and counterfeit medicines

Though reliable statistics are hard to come by, the problems posed by truly substandard and counterfeit medicines are certainly widespread, particularly where governments and their agencies are weak. Counterfeiting and substandard medicines can also proliferate when countries are undergoing a difficult transition from a centralized to a market economy, but the necessary regulatory checks and balances have not yet developed.

This is unfortunately an area in which (as noted in chapter 1) the debate has sometimes been confused because of problems of nomenclature; bona fide quality generic medicines have sometimes been misrepresented or erroneously regarded as counterfeit or substandard items. The issues are entirely distinct.

Substandard is a term applied to those medicinal products that have an inadequate standard of quality because of incompetence, negligence, or dishonesty on the part of the manufacturer (Newton and others 2002).[18] The problem exists with both originator and generic items. Although sound procurement practice can counter this problem to a large extent, constant vigilance through inspection is needed once a medicine has been procured or admitted to the market. The long-term solution must primarily lie in strengthening procurement, regulatory, and inspection systems, backed up by policing and judicial structures that ensure that regulations are truly enforced and that sanctions are imposed where necessary. Promoting the transfer of manufacturing technology from the developed to the developing world and providing technical assistance would be an effective way to counter the problem of substandard medicines quality.

Counterfeit medicines represent deliberate forgery and constitute an equally serious problem.[19] A counterfeit medicine will be produced so that both the packaging and the contents resemble the originals in their color, shape, name, and typography. Such a medicine will often be smuggled into the supply chain at some level where vigilance is lacking. Since the contents are usually of no medicinal value, such products represent a real risk to public health. Because they also present a threat to the turnover of the bona fide producer, this is an area in which recognized manufacturers and the public health authorities have sometimes worked together successfully to track down and eliminate the sources. As in the case of substandard medicines, the policing and judicial systems must provide the backing needed to ensure that the law is enforced.

For donations to be effective, recipient countries must have the capacity to manage and distribute the donated medicines

Medicines donations

The research-based industry has long been active in providing medicine donations that address priority diseases of poverty. Since 1998, 10 major companies in the Partnership for Quality Medical Donations have donated products worth $2.7 billion, which constitute only a portion of total contributions made by the industry. Merck, Pfizer, and GlaxoSmithKline have large, long-term donation programs to control and eliminate onchocerciasis (river blindness), lymphatic filariasis, and trachoma (box 2.3). Key elements of success are the effective involvement of and collaboration with governments and civil society; involvement of the community in treatment delivery; open-ended commitments of donated medicines supply; company interest in promoting learning and improvements in program implementation; and adequate administrative support and training.

A constructive development in the area of donations has been the publication of *Guidelines for Drug Donations* (WHO 1999a). The WHO led this effort, which was supported by the pharmaceutical industry. The guidelines were reviewed in 2000 and the findings show that the approach has been beneficial for recipient governments. A key aspect of the guidelines is the emphasis on donations being made only in response to recipient country requests, based on their assessments of need. A recipient should have the capacity to manage and distribute the donated medicines.

Large, well financed, and well managed global disease control programs aside, most donations at the national to local levels are short term and not sustainable for meeting ongoing medicines needs. Inappropriate donations are those that do not meet the needs of the country or that use medicines close to their expiration, which imposes the additional burden on countries of properly disposing of unusable medicines (Hogerzeil, Couper, and Gray 1997). A recent review of the effectiveness of donations by Autier and others (2002) found that inappropriate donations commonly came from small organizations with little or no field presence or experience in the pharmaceutical sector and from re-donations (often by developing countries themselves, passing on surplus items, and local in-country distributors unable to sell their medicines in the market). All of these errors are avoidable, and various attempts are being made to address them.

Even large-scale approaches have their limits. For example, the GFATM has not endorsed medicines donations, in part because the scale of need for treating AIDS, TB, and malaria is simply too great to rely on donations as a major means of medicines supplies.

Prescribing and dispensing

Inappropriate use of medicines is both wasteful and dangerous. A 1994 study conducted by the World Bank reviewed the causes of medicine waste in Africa. It suggested that for every $30 of medicines reaching the periphery, $15 could

Box 2.3

Community-directed treatment with ivermectin (Mectizan©): An example of an effective medicines donation and distribution strategy

Source: Oswald, Leontsini, and Burnham 2004.

According to WHO, onchocerciasis (river blindness) is endemic in 30 African countries. It also occurs in specific locations in six Latin American countries and in Yemen. An estimated 18 million people are infected with onchocerciasis. Among these, approximately 0.3 million persons are already blind from the disease. The recently developed and introduced community-directed treatment with annual doses of ivermectin could make it possible to largely eliminate this blinding disease burden from the affected countries in Africa and Latin America by 2010.

In 1987, Merck declared its commitment to donate ivermectin (Mectizan®) free of charge worldwide "to all that need it for as long as needed." Because it requires only an annual dose and it can be easily administered, the provision of this donation is a crucial element in efforts to control and eradicate this debilitating illness. In the mid-1990s, studies were undertaken to assess the most effective means for distributing and delivering ivermectin in the community.

Community-designed distribution systems achieved better coverage than those designed by control programs. Furthermore, they appeared to have a greater potential for sustainability, as demonstrated by a number of factors: the commitment of community leaders and distributors, the high level of community involvement and a willingness to commit available resources, the perceived benefits of ivermectin and a high demand for treatment in endemic communities, and the community's ability to determine and rectify problems within the distribution methods.

By 1997, community-directed treatment with ivermectin was adopted as the "principal method" for onchocerciasis control in Africa. The components of this protocol are as follows: the selection of the distributors; the mode of procurement and collection from the central supply; the form of communication used within the community; the method of dispensing the medicine; cost sharing; and the level of supervision and referral of adverse reactions.

The flexibility inherent within community-directed programs makes them suitable to the variety of circumstances in which treatment programs are required. The convenience of treatment, and thereby coverage, is increased by allowing the community to determine when, where, how, and by whom the medicine will be dispensed.

A review of the ivermectin distribution program in Uganda showed problems with delivery and treatment prior to the adoption of the community-based model. Studies also showed changes over time, consistent with changes to the community-directed approach, in the involvement and empowerment of women as distributors in the program. Initially, few women were chosen. Reasons cited included lack of interaction and trust among women, too much other household work, meeting sites being too far away, not being informed about meetings, sickness, restrictive husbands, and lack of information. However, when gender-responsive approaches were used in the community-directed activities, information and knowledge changed, trust increased, and more women have become distributors. Women have performed very well in this role. It has been empowering for them, in part because community-directed treatment builds on an understanding of gendered roles in providing healthcare and health-seeking decisionmaking in the community.

The success of community-directed treatment with ivermectin has been carried over into other donation and disease control and elimination programs, including the extension of ivermectin donations to treat lymphatic filariasis (in conjunction with a GlaxoSmithKline donation program for albendazole) and Pfizer's donation of azithromycin (Zithromax©) to treat trachoma. These disease control programs have produced important advances in knowledge about what approaches are successful in ensuring medicines delivery and use at the community level.

Even when a medicine reaches isolated populations, appropriate treatment may be out of reach because of a prescriber's lack of knowledge

be squandered as a result of poor prescribing, and a further $3 as a result of noncompliance by the patient (World Bank 1994).

Inappropriate use can and does also result in injury. In infectious disorders, such as malaria and TB, it can also result in a massive increase in resistance to treatment. In one study in Tanzania, 75 percent of health workers were found to be dispensing subtherapeutic doses of malaria regimens to stretch inadequate state funding, a practice that is notoriously prone to induce resistance (Mnyika and Kilewo 1991).

The WHO introduced the use of international nonproprietary names in 1950. The existence of several names for the same substance can be a source of potentially dangerous confusion; the use of a universally recognized and accessible name can reduce the confusion and potential for error. To date, more than 7,000 international nonproprietary names for generic and newly developed products have been selected, published, and translated into five languages (WHO 2004d).

Prescribing medicines

Even when a medicine reaches more isolated populations, appropriate treatment with it may be out of reach because of lack of knowledge on the part of the prescriber (inappropriate prescribing). Inappropriateness may involve, for example, over- or underdosage, the use of several medicines where one would be sufficient, or the use of an entirely unsuitable agent (Pavin and others 2003). It can also involve prescribing an expensive patented medicine despite the fact that a virtually identical generic product is available free of charge or at fraction of the price.[20] In impoverished settings, this wastes already scarce resources.

Several developing countries have attempted to promote generics prescribing as a general policy. Such campaigns are directed to prescribers and the public. As a rule, the introduction of simple prescribing guides for health workers (such as standard treatment guidelines and formularies) in which the recommended medicines are listed primarily or exclusively by generic name, has had a significant effect, especially where these generic products are readily available countrywide through a national supply system. However, in various countries (and in the urban and private sectors in most countries) resistance to this concept has been experienced and there has been relatively little impact, so that originator products often remain dominant.

Efforts to promote generics use in Nigeria, Pakistan, the Philippines, and some parts of Latin America are undermined by the deeply rooted professional and public perception that "lower-priced pharmaceutical equivalents . . . are necessarily of a quality inferior to the brand-name products sold by large, well-known firms" (Velásquez, Madrid, and Quick 1998). This failure is striking in view of the relative success of generics prescribing schemes in a range of countries, such as Denmark, Germany, the United Kingdom, and the United States. However, in these countries, very firm measures have been required to

Some excellent work has been done on examining the quality of prescribing in particular countries

change prescribing habits. It seems obvious that the successful introduction of generics prescribing requires good public relations and persuasion (if not compulsion), and not merely favorable prices.

Some excellent work has been done on examining the quality of prescribing at the country level so that corrective action can be taken where necessary (Laing 2001; Pavin and others 2003). A simple method is the ABC analysis of the medicines procured nationally, which is likely to point to certain gross faults in prescribing. The following summary is from *Drugs and Therapeutic Committees: A Practical Guide* (WHO and MSH 2004, p. 82):

> Most pharmacists and managers know that only a few drug items account for the greatest drug expenditure. Often 70–80 percent of the budget is spent on 10–20 percent of the medicines. ABC analysis is the systematic study of annual medicine consumption and cost in order to determine which items account for the greatest proportion of the budget. ABC analysis can:
>
> - Reveal high usage items for which there are lower-cost alternatives on the EML [essential medicines list] or available in the market. This information can be used to:
>   a. Choose more cost-effective alternative medicines.
>   b. Identify opportunities for therapeutic substitution.
>   c. Negotiate lower prices with suppliers.
> - Measure the degree to which actual drug consumption reflects public health needs and so identify irrational drug use, through comparing drug consumption to morbidity patterns.
> - Identify purchases for items not on the hospital or clinic essential medicines list (i.e., the use of non-formulary medicines).
>
> ABC analysis can be applied to drug consumption data over a one-year period or shorter. It can also be applied to a particular tender or set of tenders.

Other sources on country-specific prescribing practices can be found in Ph.D. theses or work conducted by bilateral aid agencies. Supplementary evidence of faults in prescribing may be found by comparing orders received from comparable districts or institutions (which may point to overconsumption or to variables such as the training of the prescriber and local and cultural preferences). Studies carried out in comparable countries can suggest common features of prescribing practices.

More sophisticated methods for the study of prescribing include establishing a prescribing and patient care survey, using WHO's healthcare facility medicine use indicators (Hogerzeil 1993, updated 1997) and country progress indicators (WHO 2000a). Where sufficient resources are available, one may establish a series of periodic medicine-use surveys, such as those carried out biennially in Zimbabwe (Trap and Lessing 1995), or monthly

Physicians in developing countries seek peer-reviewed, evidence-based data on medicines to improve their prescribing skills

self-monitoring at health centers or district level, as developed successfully in an area of Indonesia (Sunartono 1995). In all these respects, however, it is necessary to consider carefully how much investment in research is justified before proceeding to action, particularly since some of the faults in the existing situation may be entirely obvious or may have been documented in previous studies.

It is tempting at first sight to attribute much irrational prescribing to the fact that, in much of the developing world, prescribers commonly do not have full medical training. While it is beyond doubt that basic training is commonly inadequate, this does not sufficiently explain the fact that remedies are poorly selected and applied. At the primary care level, the number of essential medicines likely to be available and in regular use is quite small. So it is not an impossible task to provide prescribers who have a basic education in nursing or as medical assistants with sufficient guidance to diagnose the most common conditions likely to be encountered and to prescribe with a reasonable degree of competence. Fully qualified prescribers may prove to prescribe irrationally, especially where their prescribing is linked to an income from dispensing, which creates a temptation to overprescribe as a means of increasing earnings (Trap, Holme Hansen, and Hogerzeil 2002).

In some countries, there is intensive commercial persuasion to prescribe newer and more expensive remedies even in situations where they offer no advantages over older, much cheaper products. However, there are evident risks in aggressive promotion in an environment where educational standards are low, objective sources of data are hard to come by, and no resistance to advertising has developed.

Physicians in developing countries actively seek peer-reviewed, evidence-based data on medicines to improve their prescribing. Improving this situation as a matter of policy will provide valuable assistance to physicians in developing countries (Hafeez and Mirza 1999).

Guidelines on the relationships between members of the medical profession and representatives of the pharmaceutical industries are currently the focus of recommendations—in some cases regulation—by professional organizations. In a brief review that examined routine professional exchanges of goods and services from industry representatives to physicians in both developed and developing countries, Ann McGuaran (2002) noted an increasing trend toward regulation of such exchanges by physician organizations to ensure their professional appropriateness. The Royal College of Physicians (United Kingdom) recently issued updated guidelines that covered a wide range of situations, including the following:

- No conditions should be attached to gifts, items of equipment, or aid.
- Under no circumstances should cash or objects in kind be accepted by individual physicians, and gifts, honorariums, or hospitality received must be declared.

Improving prescribing will require both short- and long-term efforts

- Speakers at company-funded meetings should not be chosen solely by the firms, and the hospitality firms provide at meetings with an educational purpose should be modest.
- The payment of reasonable expenses and honorariums is acceptable for larger and overseas meetings but should be handled through the independent scientific body and not paid directly to individual physicians.
- All research must be cleared by the doctor's research ethics committee, and all financial matters should be managed by institutional finance departments (RCP 2004).

Realistic and cost-effective approaches for developing rational prescribing according to WHO standards have been documented (Laing, Hogerzeil, and Ross-Degnan 2001). Not all have been tested under strictly controlled conditions, but the teaching methods tested over a period of years at Groningen University and McMaster University, partially summarized in a handbook available on the Internet, have a creditable record of success and have, through the medium of WHO, been widely adapted for use elsewhere (De Vries and others 1995).[21]

Improving prescribing will require both short- and long-term efforts. Recognized approaches are documented in the literature and a considerable fund of experience is available through the International Network for Rational Use of Drugs, which works in many developing countries as well as organizes training courses (INRUD 2002). Short-term methods include the development of national standard treatment guidelines (often based on the numerous established handbooks of this type), formularies, and bulletins. There are many excellent publications of this type that can be adapted to national needs; for example, the International Society of Drug Bulletins is a valuable source of advice, support, and draft texts. Some well known textbooks and reference volumes, such as the *Merck Manual* (consumer and professional editions) are provided throughout the developing world (in local languages) by the International Pharmaceutical Federation Pharmabridge Program and the International Council of Nurses trunk library program.

In hospitals and other institutions, therapeutic committees, with the full participation of medical and pharmaceutical staff, are capable of setting and maintaining high standards, in order to both improve patient care and economize on resources. The reorientation of the pharmaceutical profession can provide a promising new resource working toward the better use of medicines.

When seeking to influence prescriber behavior, it is essential to use means that will not be resented by those concerned. Physicians, in particular, are most likely to be responsive to efforts to improve their standards of practice if they emanate, at least in part, from within their own profession instead of being imposed upon them. The creation of therapeutics committees, noted above, is an important step in this direction. A national medical association and a nursing association should participate in developing these approaches and should

The retail pharmacist, dispenser, or drug seller is often the patient's primary source of information about medicines

be encouraged progressively to assume responsibility for them. When medical training is extended and upgraded, efforts must also be made to encourage the development of the professional's communication skills as well as his or her technical abilities. In particular there is a need to develop skills in communicating with patients, including the provision of facts and advice relating to medicines.

Finally, the issue of advertising and promotion for medicines will need to be tackled, generally in line with international ethical standards but also taking into account the particular susceptibility to commercial persuasion that a population may have when it is exposed to these influences for the first time.

Dispensing medicines

Much of what has been said above regarding the prescriber applies by analogy to the retail pharmacist, dispenser, or drug seller. The drugstore, whatever its nature, is often the patient's primary source of information about medicines, and in matters of self-medication it is likely to be the only one. Fully trained professional pharmacists will not be available for retail duties in all countries in the foreseeable future. But much can be done through basic training and follow-up documentation to ensure that the retailer, whatever his or her level of education, provides reasonable assistance to the purchaser of a medicine to understand its nature and how to use it appropriately. In some cases, trained pharmacists may be available but are not fully utilized in a manner consistent with their level of training (Professor R.R. Chaudhury, personal communication, 2004).

Use of medicines in the home

A medicine is not always used as instructed by the prescriber or indicated on the package. The verbal instructions may not have been clear or they may have been misunderstood. The text on the package may be in a foreign language or the patient may be illiterate. Common misunderstandings ("two doses are better than one") may call for correction. Gender issues can also play a role in determining the extent to which medicines are used appropriately in the family. It is often the woman who brings home the medicines and administers them. Women must thus have sufficient knowledge to select those medicines that are available without prescription, and to ensure that the medicines entering the home are used appropriately. Inequality in educational systems typically is to the disadvantage of women and girls. As in the case of health professionals, it is necessary to envisage both short-term and long-term approaches among the general public in order to promote the rational use of medicines.

Short-term approaches can be constructed around the fact that irrational use commonly reflects popular misunderstandings regarding the nature and use of medicines. Poster campaigns in clinics and brief messages transmitted by radio and TV have proved effective in correcting some of these misconceptions.

Irrational use commonly reflects popular misunderstandings of the nature and use of medicines

They can also encourage the consumer to seek the advice and information that he or she needs to understand the proper use of a medicine, especially where no written information is accessible or the user is illiterate. In the longer term, the main solution must lie in improved standards of education, particularly where there is a need to correct a lag in progress in educating girls and women. The proper use of medicines should be a component of popular health education.

Promoting Rational Drug Use (2000) is an online compilation of course materials and references available on the Boston University website.[22] Included in this collection is a session on "Effective Community Education" by Professor Anita Hardon that describes a systematic approach to providing information to patients and to correcting widely held misconceptions (detailed information is available online[23]):

Step one: Investigate

The investigatory stage is essential and at the core of the communication process. It should address the following issues:

- What is already known about the problem?
- What new kinds of information are needed?
- What are the characteristics of the target audience?
- What development communication resources are there?
- How should data be generated? (both quantitative and qualitative)

Step two: Plan communication activities

Step three: Develop communication materials

Step four: Test and revise materials

Step five: Implementation

Step six: Monitor, evaluate and revise

These efforts must be complemented by others, especially those concerned with the information provided to the consumer at the point of sale. Both pharmacists and other medicine sellers need to become accustomed to providing information and advice to all customers purchasing medicines, whether over the counter or by prescription. Finally, as noted earlier, there needs to be an ongoing effort to involve the public and its representatives in developing better standards of medicine use and improving the accessibility of information and advice to the individual patient.

Some preconditions for success

Although increasing access to medicines should be high on the agenda of the world community, it competes with many other priorities, not least of which are international conflicts and the problem of terrorism. Only in a very exceptional situation, notably the global AIDS epidemic, does the issue of medicines come sufficiently to the fore to gain a fair measure of social and political support at the international level. At the same time, situations of armed conflict, whether national or regional, can seriously disrupt the supply of medicines.

Global preconditions

One major precondition for continuing and expanding global efforts to develop access to medicines is the attainment of a greater measure of international peace and security.

Another prominent precondition is the state of the global economy. Despite the evidence that improved access to healthcare can be a significant stimulus to economic development, paying for itself many times over, investment in health is still not sufficiently accepted at global or national levels as a significant tool for promoting economic growth and social welfare when resources are limited and other priorities present themselves. On the contrary, when the economy stagnates or contracts, it is common to see health budgets reduced. It is not clear whether economic recession has been responsible for the failure to meet international commitments to provide assistance, with some of the world's wealthiest countries remaining farthest from achieving their long-standing commitment to the UN development aid target of 0.7 percent of GDP. It may provide a partial explanation—or at least a plausible excuse.

A balanced national medicines policy is a necessary component of an overall health policy

National preconditions

National health policies and systems are not always fully attuned to ensuring that medicines are available, affordable, or appropriate:

- Expanding access to essential medicines requires attention to a diverse set of policy challenges. Solutions must begin with an understanding of local health conditions in their broadest epidemiological, economic, regulatory, and cultural contexts.
- Improvement in access to essential medicines in not an event but a process that requires ongoing support from a range of stakeholders, beginning with the government, but also including the private sector.
- Reforms are most effective when they focus on the most critical access problems rather than attempting to address all barriers simultaneously.
- As a corollary to the first and third points, emphasis on data collection and analysis is required to ensure accurate identification of the most critical problems and barriers (WHO and IFPMA 2000).

A number of basic structural problems that underlie the barriers to access are often encountered. Some basic needs must be met if access to medicines is to be assured. These needs have been well documented in the literature and in national and consultant reports and are briefly outlined below.

National medicines policy

A balanced national medicines policy is a necessary component of an overall health policy; this applies to all countries. Without this balance, it is more likely that stop-gap, uncoordinated solutions will be employed, with little prospect of long-term improvement. The policy will need to be comprehensive and well planned; it will have to be based on well proven general principles, including the essential medicines concept; it will have to be adapted to national needs;

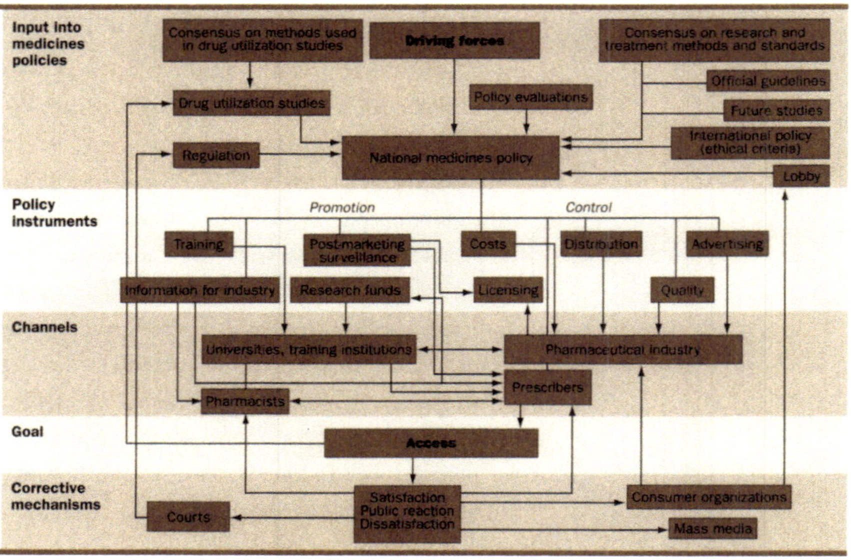

Figure 3.1
Theoretical model of a national medicines policy

Source: Dukes 2003.

The current weakness of national systems in supply and use of medicines is in part a question of lack of appropriately skilled human resources

and its implementation will need to be monitored (WHO 2002c). In many countries, making marginalized or underprivileged groups a priority will be a distinct component of policy. There are striking examples of broad national medicines policies, particularly those incorporating the essential medicines concept, with concrete achievements to their credit, though success on one front may go hand in hand with relative failure on another.[1, 2]

Economic situation, setting priorities, and political will

There has been increasing global awareness of the link between medicines, health, and economic growth. However, this awareness is frequently not reflected in practical terms on the international economic policy level. For example, the International Monetary Fund (IMF) continues to recommend (as a condition of loans) that inflation in many developing countries be controlled by applying limits on health sector spending. The movement to abolish school and health service user fees has increased lately, with demonstrable benefit for the poorest people. However, with continued imposition of spending limits, services and human resources cannot be expanded to meet increased demand.

A weak national economy will not be rendered strong overnight, but if there is a realistic plan for economic development, a developing country will have a much greater ability to attract both donor funding and loans for healthcare and medicines supply.[3] Poverty reduction strategies can effectively and explicitly make the link between health and economic growth by detailing the human and economic cost of preventable and treatable diseases, identifying constraints and obstacles that prevent access to affordable (or in the case of very poor people, free) services and medicines, and explicitly developing and applying strategies to remove these constraints and advance expansion of services.

Human resources

Though the current weakness of national systems in supply and use of medicines is partially a question of lack of finance, the lack of appropriately skilled human resources is at least as critical. Basic schooling, further education, and cross-training of health workers all leave much to be desired throughout much of the developing world, and correction is needed at all levels. Some countries also suffer severely from the emigration of health workers and the inability to attract or retain competent workers from abroad. From procurement to rational use, the system depends heavily on staff who are properly trained and sufficiently motivated and rewarded to ensure that they are retained and perform conscientiously. Corruption, inefficiency, and managerial weakness in the public sector will be effectively relieved only when the government service is sufficiently well financed. Here, as in other fields, priorities have to be set if limited resources are to be used to the best advantage, and unorthodox solutions may need to be applied.[4]

The working group recommends that wealthy countries adopt voluntary restrictions on recruiting developing country health workers

One study estimates that approximately 20,000 health professionals emigrate from African countries annually (Raufu 2002). UNCTAD estimates that each migrating African professional represents a loss of $184,000 to Africa. Paradoxically, Africa spends $4 billion a year on the salaries of 100,000 foreign experts (Pang, Lansang, and Haines 2002).

Stilwell and Awofeso (2004) elaborated a focused discussion of the issues surrounding "brain drain" in African countries (specifically Nigeria). The causes for outmigration include the following factors:

- Doctors trained to levels that are more advanced than required by local health realities.
- Poor remuneration, with huge salary discrepancies between local wage scales and developed country compensation models.
- Lack of incentives for overseas-based doctors to return to their home countries.

Suggested management strategies to reverse these trends include the following actions:

- Establish full fee–paying private universities to train doctors for export.
- Intensify training of allied medical staff whose skills and competencies are suitable for current healthcare needs, particularly in rural areas.
- Increase public sector salaries and provide additional support, such as subsidized housing and transportation.
- Provide a stimulating environment for intellectual growth, including access to computers, the Internet, and journals.

While also citing poor remuneration, another study widened the scope of reasons for the outflow:

> [B]ad working conditions, an oppressive political climate, persecution of intellectuals, and discrimination. Researchers cite lack of funding, poor facilities, limited career structures, and poor intellectual stimulation as important reasons for dissatisfaction. Other key reasons for emigrating are personal ones. These include security, the threat of violence, and the wish to provide a good education for their children. (Pang, Lansang, and Haines 2002, p. 499)

Bilateral agreements to address the ethical issues of recruitment practices by developed countries, compensation to be paid by departing professionals, and flexible job requirements that would permit public sector employees to simultaneously work in the private sector are among other strategies to improve health worker development and retention in poor countries (Pang, Lansang, and Haines 2002).

Responding to the increasing crisis of outmigration of skilled health workers, the Working Group on Access to Essential Medicines recommends that wealthy countries adopt voluntary restrictions on active recruiting of

Currently, the most useful role of the retail pharmacist is to provide information and advice to patients

developing country health workers and reimburse developing countries' training costs, under certain agreed conditions, for workers who immigrate.

Scaling up human resources in the health sector will require decades of focused attention to improve educational opportunity, including targeted strategies to address equality of access for women and ethnic minorities. Complete national coverage by professionals trained to full academic standards cannot be attained without years of sustained effort.

Experience to date in the field of human resources within the pharmaceutical sector can be summarized under three main headings:

Managers. Many development projects have found it possible to train competent managerial staff in areas such as finance, procurement, supply management, and distribution partly through diploma-level courses and on-the-job training by working for a period alongside local and expatriate experts. Key managers may benefit from training periods abroad, especially in other developing countries where analogous situations have been tackled successfully.

Prescribers. Given an adequate information flow, a great deal of the clinical management of common disorders, including diagnosis and medication, can be competently handled by staff trained to diploma level (2–3 years of training) who are carefully supervised and well supported with continuing education. Fully trained physicians can then handle cases referred to them from primary care.

Pharmacists and pharmacy technicians. The number of trained pharmacists and pharmacy technicians, especially outside of main urban centers, remains woefully inadequate. A small number of pharmacists play a valuable role in medicines regulatory agencies, inspectorates, hospitals, and medicines manufacturing units, but the remainder generally operate urban retail pharmacies that typically serve small affluent groups. To date, no developing country has been successful in meeting the aggregate numbers of trained professionals needed, with even less success in attracting them to rural areas. Currently, the most useful role of the retail pharmacist is to provide information and advice to patients. Given the absolute lack of pharmacists in developing countries, innovative strategies to train lower-level providers, such as pharmacy assistants (dispensers, technicians) will be needed. In many cases, village-level shopkeepers can receive basic training to dispense some essential medicines, under the oversight of pharmacists.

In the meantime, the reorientation of academic pharmacy training for more specialized tasks (such as medicines policy, inspection, and manufacturing), which has proceeded favorably in various industrialized countries, appears to be taking root more widely in the world. In most communities, it is possible to identify resource persons who have influence and enjoy authority among the population. Much can be done to develop their role, support their activities, and enlist their help both in designing and in implementing policies in the field of medicines.

To create the various forms of expertise nationally, donors need to have a major ongoing training component that is fully institutionalized; otherwise a long-term situation of dependence may result. Short-term training courses are helpful but are not an adequate substitute. A network of international agreements is also needed to develop collaboration between countries on the migration of health workers, in order to encourage health workers to seek employment where they are most needed and to remove purely bureaucratic obstacles to regional and internal migration. There is a need for innovative solutions and strategies to effectively address the critical shortage of human resources (box 3.1 shows one example).

Information services

Access to reliable information on medicines and their appropriate use has proved essential to rational prescribing and consumption in every country. Public involvement in this process is particularly necessary in countries where professional training is limited, though in fact, most countries in the world now benefit from the provision of impartial or officially sanctioned prescribing information services. A great deal has often been achieved by relatively simple and low-cost methods. One of these is the production of simple therapeutic guides or standard treatment guidelines for prescribers, which set out

Box 3.1

Developing innovative responses to the need for human resource expansion

Source: International Dispensary Association Solutions.

The International Dispensary Association, through its recently launched consultancy arm IDA Solutions, is proposing an innovative response to the issues surrounding human resource training and capacity building. Based in South Africa, the overall objectives of the project include:

- To provide a buffer stock of critical antiretrovirals and supplies that can be mobilized within 24 hours and reach 80 percent of African destinations within 5 working days.
- To train and build capacity in Africa in the proper selection, quantification, procurement, storage, and distribution of AIDS-related medicines and medical supplies.

The project will start with a focus on the supply of first-line antiretrovirals and diagnostics by providing technical support in the form of organizing the actual supply (including estimating needs and ensuring a buffer stock) and global tools to be made available to all, with targeted assistance provided in priority countries. Where it is requested and where it is offered as is a comparative advantage over existing services, including the supply from its buffer stock, the project could facilitate procurement. Direct procurement will be considered only when it can offer distinct advantages in price, labeling, or other contract terms.

Scheduled to launch in early 2005, the project will be organized and run as a nonprofit operation with the operational objective of using a small staff of multiskilled experts to support local projects while creating a network of local consultants in countries throughout Africa.

Frequently, horizontal and vertical programs exist alongside each other, at times resulting in duplication in one area and gaps in another

the means of diagnosing and treating the disorders and symptoms most likely to be encountered, for example, in a rural health center. Such guides, well edited and in pocket format, have (in combination with other measures) had a considerable effect in improving the quality of medicines use in the countries where they have been introduced.[5]

An acceptable variant on this is a national formulary in which similar information is provided but it is presented in relation to medicines rather than to disorders. The availability of well proven national models and of a recent model formulary from WHO (WHO 2003e) mean that the production of new prescribing guides or national formularies is today neither difficult nor costly. A complementary approach that appears similarly promising is the production and dissemination to health workers of national medicines bulletins at regular intervals. Such bulletins can update the therapeutic guides and formularies where necessary, but their primary purpose is to alert professionals to current problems relating to disease and the use of medicines, and suggest ways to avoid errors and misunderstandings. The International Society of Drug Bulletins has given valuable support in the form of draft texts and advice when new bulletins are established. Like the prescribing guides and formularies, such bulletins have been shown to be capable of promoting rational use of medicines to an extent that is entirely disproportionate to the very low expense involved in their production.

With much, if not all, of this information available and constantly updated on the Internet, incorporating technology (such as computers, reliable sources of power, satellite access to the Internet, and training of medical personnel in computer literacy) into long-term planning health sector strategies could yield promising results in supporting continuing education, as well as linking isolated clinicians and other health providers (Fraser and others 2004).

Horizontal linkages

An empowered ministry of health could have a significant impact on critical medicines issues. For example, an increased ability to influence the policies of other ministries, especially those dealing with import duties and taxation, could move toward resolution of these issues when they adversely affect the flow of medicines. Other related advocacy roles could include strategic planning in partnership with medical and pharmacological training facilities and ongoing interaction with popular media sources to disseminate information.

Donor coordination

Frequently, horizontal and vertical programs exist alongside each other, at times resulting in duplication in one area and gaps in another. A number of governments have also preferred to negotiate with donors individually on medicines issues, apparently hoping in this way to obtain a larger volume of support. Bilateral donors and development banks commonly succeed in establishing

Trust in health services and in the providers of health services is a fundamental component of effective community participation

mutual links so that they can coordinate their efforts, and this coordination appears to be optimal where there is a broad and open development program.

Community participation

Advocates of community participation, as well as the wider development community, consider the full involvement of a community to be a key way to improve governance, including the equitable allocation and utilization of resources in the health sector. Community participation can provide a mechanism through which potential beneficiaries of health services become involved in the design, implementation, and evaluation of activities, with the overall aim of increasing the responsiveness, sustainability, and efficiency of health services or health initiatives and programs.

Trust in health services and in the providers of health services is a fundamental component of effective community participation. Part of the tension between traditional and biomedical systems can be understood through an analysis of patient confidence and the degree of cultural comfort with the treatment models. This in turn speaks to local beliefs and community traditions, both of which are inherent features of indigenous systems. Biomedical practice, with its positivist tradition, frequently continues to be less responsive to these issues than indigenous systems. However, underlying this issue of confidence are the economic contexts in which this health-seeking behavior occurs. Trust is built through the effective and timely response to need. Clinics and health posts with minimal or no reliable stocks of essential medicines consistently fail, through no fault of the healthcare providers, to inspire confidence when patients cannot obtain the medicines needed to treat their complaint or when patients are provided with a prescription and referred to a private facility or a pharmacy where the cost of the medicines is beyond their ability to pay.

In a paper commissioned by the working group addressing the role of the community in improving access to medicines, Oswald, Leontsini, and Burnham (2004, p. 6) discuss concepts of inclusion, representation, and participation:

> One area of concern that Zakus and Lysack raised was the true form of representation within community participation, in regards to both the receipt of health services and the widely touted social benefits of such programs, and it drew attention to "who it is that is included in the community (and thus community participation) and who is not," when they questioned whether "empowerment and health promotion sufficiently challenge power structures that systematically operate to leave some people in poorer health than others?" (1998). Barbara Klugman recently drew attention to the idea that the "mechanisms for local-level community participation in health tend to exclude women," and that such "mechanisms are also generally administrative, rather than decision-making, regarding health or medicine priorities" (Irwin and Ombaka 2003). Similarly, Vlassoff and Moreno commented that

Use of paid community health workers in distributing essential medicines to poor and isolated populations is a promising model

community participation, in the form of primary health care, was often promoted without adequate consideration of gender (2002). On the other hand, despite this neglect and the growing problem of income inequality over the last decade, Moss pointed out the recent worldwide adoption of a gender perspective in health and development research, along with new efforts to protect women's rights and target discrimination (2002). At the turn of the 21st century, women and children suffered poverty more than men, indicating the need for truly participatory methods to deliver health services to the community for the alleviation of women's poverty and empowering women to combat discriminatory practices (WHO 1999; WHO/TDR 2003).

A commitment to ensuring access to essential health technologies must focus on the needs of communities living in poverty. However, the gendered structure of poverty—in addition to the gender dimensions that facilitate or inhibit access—are a critical component of this issue and require an equal, if not a greater, consideration in program orientation. The aim of the working group is to "ensure that poor people in developing countries obtain access to a reliable supply of high-quality, affordable medicines for the diseases that threaten their lives, undermine their communities, and enmesh their countries in want and despair" (Irwin and Ombaka 2003). Full community participation in interventions to increase access to essential health technologies holds great promise to be both a sustainable and pro-poor approach. Success, however, will require a greater consideration for the role of women. Moss stated, however, that women are increasingly taking public roles in their communities that are integrated with their household roles (2002). Two examples, from Ethiopia and Burkina Faso, demonstrate the promise of such approaches in achieving significant strides toward the control of malaria within a gender-sensitive approach that seeks to improve the management of disease in the home.

The empowerment of women in many parts of the developing world will require focused effort to achieve what is a profoundly transformative process within societies. This cannot be achieved without years of consistent, comprehensive effort that crosses all segments of societies. Examples of the outcomes of these efforts include the expanded education of both girls and boys, unrestricted access of women and girls to health services, the restructuring of property and inheritance rights, the prevention of domestic violence, and the participation of women at all levels of governance.

Community health education is an arena in which the engagement of community members in designing and implementing programs holds great promise. Culturally appropriate educational campaigns that incorporate local dialect and symbolism deliver the message in a targeted, effective way. Understanding the constraints imposed on the use of medicines by largely illiterate populations requires focused attention to nonwritten instructions on dosage,

Ongoing feedback at all levels will enable successes to be reinforced and replicated

timing, and route of administration. These efforts are time-consuming but essential. Engaging and training paid community health workers are effective ways to meet these needs.

The use of paid community health workers in distributing essential medicines to poor, widely dispersed, and isolated populations is a promising model. For example, Haiti, the poorest country in the western hemisphere, has a thriving program of antiretroviral delivery to rural, frequently isolated AIDS patients that has continued uninterrupted through natural and manmade disasters (Mukherjee and others 2003).

Oswald, Leontsini, and Burnham (2004, p. 10) reviewed some of the existing literature on community health workers:

> According to Walt (1990), some have argued that enhancing the role of community health workers to be curative agents within the community is dangerous, as it may lead to inflated claims of ability, akin to "mini doctors," that are not substantiated due to poor training. The widespread threat of drug resistance combined with the improper use of medicines adds considerably to the strength of this argument. Walt concluded, however, that community health workers cannot establish their credibility within a community without being able to offer access to drugs or curative treatments. Without being able to offer this access, their preventive work is undermined.

The authors found that some studies suggest that a sustained supply of essential medicines (understood as curative interventions) is critical to maintain the confidence of community members in both the local health system and individual community health workers (see Stone 1986; Van der Geest 1992). Others acknowledge the importance of sustained medicine availability but maintain that it must also be accompanied by behavioral changes (see Delacollette and others 1996).

Monitoring

Ongoing feedback from the system at all levels will enable successes to be reinforced and replicated and failures to access to be identified and corrected. An independent inspectorate—capable not only of identifying problems but also of helping to correct them—is an important tool to this end. The inspectors should ideally be senior figures with experience in the area that they monitor, enjoying sufficient respect for their criticisms to be taken seriously and their proposals for change to be respected and adopted.

Institutional structures

Successful implementation of national medicines policies is found to rely heavily on the existence of a series of reputable and stable institutions with reliable financing. Alongside the bodies directly concerned with medicines policy and

A well organized consumer movement and patient organizations can contribute much to the evolution and monitoring of policy

regulation, academic institutions in such fields as medicine, pharmacy, law, and economics; professional organizations; institutions able to speak for trade and commercial interests; and financial bodies handling health insurance or reimbursement systems will all serve an important role in designing and implementing sustainable access to medicines for all the citizens of a country. A well organized consumer movement and patient organizations can contribute much to the evolution and monitoring of policy.

At the national level there must be the necessary political will to create and maintain these institutions and to provide them with the necessary finance, staff, and authority to work effectively. A proportion of donor aid must always be devoted to improving infrastructure, without which direct aid is unlikely to be optimally used.[6]

A problem for many countries is the weakness of the legal system and the judiciary. Offences are not efficiently prosecuted, and where corruption exists it is frequently ignored. In a number of countries, bilateral donors are now providing support to strengthen the legal systems, the judiciary, and the police. Again, there will have to be progressive, incremental change in social behavior and expectations before problems of this type are effectively addressed.

Civil stability

In a stable civil situation the supply of medicines can evolve progressively as funding, infrastructure, and experience permit. Medicines access has frequently been disrupted by war and insurrection, and in some instances obstacles to progress have been raised by those intolerant of change. In such situations, emergency aid in the form of donated or heavily subsidized medicinal supplies may, for a considerable time, be the only means of maintaining even a modicum of access.

Throughout the world, millions of people have been displaced due to conflicts and natural disasters. These populations are especially vulnerable to disease and injury. Many displacement victims reside for years in temporary relocation camps that lack secure food supplies, basic sanitation, or health services of any kind. Often caught between warring factions, they remain in a state of legal limbo without recourse to civil or human rights. Their plight requires focused intervention from the international community.

Summary of recommendations and concluding remarks

Expanding access to essential medicines requires attention to a diverse set of policy challenges. National health policies and systems are not always fully attuned to ensuring that medicines are available, affordable, or appropriate. Solutions must begin with an understanding of local health conditions in their broadest epidemiological, economic, regulatory, and even cultural context. Increasing access must be seen as a process requiring ongoing support from a range of stakeholders. Reforms are most effective when they focus on the most critical access problems, rather than attempting to address all barriers simultaneously. Countries need adequate data collection and analysis to assess and set priorities in problem areas.

Access to medicines cannot be addressed in isolation either from the rest of the health system or from the overall health situation in a given country. Access to medicines is not an issue that exists in a vacuum: it is an integral part of healthcare, the various components of which are mutually supportive. Measures in all these areas will need to be backed by the systematic and ongoing assessment of the needs of a particular country or population. On all levels there will be a need for institutional development and a sustainable expansion of human resources. Although the access to medicines issue ultimately is a global one, the working group, in keeping with the task force's mandate, has focused its main efforts on addressing how to increase access to medicines for the poor and in developing countries that have the greatest need for concerted, coordinated, and effective mobilization of resources to break the cycle of ill health, poverty, and declining economies.

The working group addressed its recommendations at two main levels: national and international. Especially at the national level, an attempt was made to be as operational as possible. This division into national and international

Equity of access should be a cornerstone in thinking and policymaking

levels, however, should be viewed with caution, since increasing access will ultimately involve a complex interplay of many actors operating at many levels concurrently and dynamically.

General principles

The working group found that certain basic principles underpinned approaches to the issues of and solutions for increasing access to medicines. These general principles include the human right to health codified in the UN Declaration of Human Rights (UN 1948); the right to treatment codified in Article 12 of the International Covenant on Economic, Social, and Cultural Rights (ICESCR) (UNHCHR 1966), which was clarified in 2000 to include the right to essential medicines (WHO 2002a; Hogerzeil 2003); and the right to medical treatment, including access to medicines, found in the International Guidelines on HIV/AIDS and Human Rights (OHCHR/UNAIDS 2002). However, the enforcement of these rights is not evident in the current global situation, where entire populations, particularly the poor and underprivileged, commonly have little or no access to essential medicines.

The working group also found that women's inequality and gender disparities contribute to institutionalized inequalities within educational and health systems. These inequalities limit women's and girls' access to healthcare and to needed medicines more than men's and boys'. Profound, incremental, and societywide changes must occur to eliminate these forms of discrimination. Health systems will need to be strengthened to deliver quality essential services while maintaining equity of access. Equity of access should be a cornerstone in thinking and policymaking. Simply put, in any program, the most marginalized people should receive healthcare and services at the same or greater rate than the more economically franchised. In the case of AIDS treatment, an equitable approach would target populations that live in the most resource-challenged areas first. An equitable approach to pro-poor healthcare, would be, by definition, a bottom-up approach.

The working group, in general, also recognized the need to find new ways for the main actors involved in the supply of pharmaceuticals to interact to ensure that needed medicines are available. Indeed, the discourse on the means to ensure supply was vibrant and robust. The working group agreed on the fundamental point that market competition is an essential driver for innovation, supply, and affordable prices.

Improving the availability of medicines

Availability of medicines is affected by many factors. The main ones that need to be tackled include ensuring that needed medicines are developed and brought to market and that supply and distribution systems are adequate to deliver them to the people who need them.

The WHO should take a leading role in promoting R&D that meets the public health priorities of developing countries

Improving the rate and relevance of innovation

Treating priority diseases of the poor is greatly hindered by a fundamental problem: the medicines required for some of the diseases and illnesses most prevalent in developing countries do not exist because of a lack of therapeutic innovation (MSF 2001). Another critical need is for new medicines to supplement or replace those to which microorganisms have become resistant, as is notably the case for malaria and TB.

A reorientation of medicines research, better attuned to the needs of the world as a whole, is necessary. This will require creative new research, development, and financing mechanisms; for example, the WHO Commission on Intellectual Property Rights, Innovation, and Public Health could examine alternative international models to the current patent-based system for priority setting and financing health R&D.

The for-profit private sector is not going to take up needed innovation for major infectious diseases of poverty without major involvement and subsidy from the public sector and an appropriate and supportive policy environment. The public sector is also going to have to remain a vital force. An equitable approach globally would have these innovation costs borne primarily by the nations with the broadest shoulders, such as heavily industrialized countries with strong economies that are capable of sustaining relatively high prices for the medicines they require.

Public-private initiatives, such as MMV, IAVI, and GATB, appear to be offering useful models for new medicines and vaccines development, and they should be supported. However, questions remain about governance, adequate participation by experts from affected countries, and adequate focus on priority medicines for the poor. These aspects need to be monitored, and effectiveness and best practices need to be assessed.

Successful innovation to help meet the Goals will require greater cooperation among all sectors (such as the public and private sectors, academia, foundations, and the United Nations), substantially more financing from multiple sources, clear priorities for research efforts, effective management, and technology and knowledge transfer. The WHO should take a leading role in promoting R&D that meets the public health priorities of developing countries. Medicines regulatory process reforms and harmonization need to better reflect and serve the needs of developing countries. Traditional knowledge and medicines continue to be marginalized, to the detriment of consumers. Vigilance surrounding all aspects of pharmacological practice in developing countries needs to be strengthened.

All of these issues point to the need for substantial change, which will take considerable time to implement and produce results. Taking new steps must start now.

At the national level:

- Governments should determine priorities in medicinal innovation in accordance with the most basic and unfulfilled needs of their populations,

Governments should determine priorities in medicinal innovation in accordance with the most basic and unfulfilled needs of their populations

and bring these priorities to the fore both in their domestic policies (such as through their national medicines policies, essential medicines lists, procurement strategies and budgets, and public R&D policy priority setting) and in the global forums in which they participate.

- Developing countries should be more confident about negotiating for technology transfer and more national capacity building to participate directly in R&D. Examples of innovative approaches include the DNDi approach to partnering with research institutes in developing countries, the cooperative effort between the Universities of Nairobi and Oxford on AIDS vaccines trials; the Kenya Medical Research Institute's partnering with GlaxoSmithKline and the University of Liverpool on the development of a new antimalarial; the Merck Vaccine Network Africa training center at Moi University in Kenya; Merck's partnership with the Harvard AIDS Program in the Enhancing Care Initiative to build infrastructure for vaccine delivery; and the Pfizer partnership with Makerere University in Uganda and the University of Utah. Even in countries with very limited resources, some steps can and must be taken to formulate a national research policy and provide the funding and infrastructure needed to implement it, either independently or in collaboration with foreign, regional, or global institutions.
- The regulatory environment should reward sound research into priority diseases. For example, a country could devise a fast-track system for priority medicines, based on national health priorities.

At the international level:

- Public investment in medicinal research should be expanded to meet the most pressing needs of developing countries and poor populations, including developing knowledge based on indigenous medicines. The international community should not rely on the research-based industry to be the primary vehicle for developing medicines needed in developing regions. Certainly they need incentives and constructive structures to participate according to their comparative advantage. But new ways of approaching innovation, such as Folb has described in detail (Folb 2004), should be considered and pursued with some urgency. The WHO Commission on Intellectual Property Rights, Innovation, and Public Health should examine alternative international models to the current patent based system for priority setting and financing of health R&D. Recent papers commissioned by DFID also support the value of taking new approaches to technology transfer, patent regimes, intellectual property management, and local production as ways to meet the demands for increased access to medicines (see, for example, Hill and Johnson 2004; Lewis-Lettington and Banda 2004).
- International standards for ethical research, such as those elaborated by the Declaration of Helsinki, should be applied in all countries.

Each country should develop a list of essential medicines as a basis for procurement and supply decisions

Developing more reliable procurement and supply systems

Many national procurement and supply systems for medicines, whether public or private, are inefficient or poorly attuned to current needs. Procurement is not always in line with what is needed, funds are not optimally used, and medicines are commonly out of stock in both urban and rural areas. Procurement and supply systems in developing countries need to be more effective and reliable, making the best possible use of public, private, and nonprofit channels, and ensuring that a reliable supply system is extended to rural areas. Each country should develop and keep updated a list of essential medicines that reflects its priority health needs and that is used as a basis for procurement and supply decisions.

At the national level:

- All potentially efficient systems for procuring and distributing medicines supplies, whether public, private, or maintained by NGOs, should be encouraged and assisted to develop. This will require country-level, ongoing capacity building. Low-income countries especially need ongoing technical assistance to build expertise in effective procurement, quality control, and quality assurance systems. National regulatory bodies urgently need strengthening in developing countries, and the judicial system should be provided with the human and material resources to enforce these regulations and eliminate abuses that can lead to waste and loss.
- The advent in recent years of the GFATM and the World Bank MAPs provide developing countries with valuable resources and incentives to improve their procurement and medicines management systems. Both organizations promote an assured-quality and lowest-price approach. The GFATM asks recipient countries that receive funds for medicines purchases to demonstrate that they have a competent national system for selection, procurement, quality assurance, supply, and distribution. Initial concerns that the fund would prompt parallel procurement and supply systems are being allayed. The fund has emphasized that national systems should be strengthened, not replaced nor sidelined. The World Bank published a detailed technical guide in early 2004 that should be very helpful at the country level in addressing these systems challenges (World Bank 2004a).
- The WHO prequalification project to identify good-quality products for HIV, TB, and malaria medicines for procurement by UN agencies is also helping low-income countries that have very limited quality assurance capacities to improve procurement by providing key quality indicators for suppliers and products. The working group endorses the WHO prequalification approach for use by countries and supports its expansion.
- Pooled procurement schemes remain a tantalizing, yet still underused, avenue for improved procurement. No one model for pooled procurement

Substandard medicines present a serious problem, especially in developing countries with limited regulatory and enforcement capacities

exists. The degree of cooperation and shared or combined systems depend on the participants, local and regional characteristics, and purchasing needs. Other examples of pooled procurements include disease-specific international initiatives such as the GDF and MMV. All of these strategies should be explored by developing countries. Countries pursuing this strategy should take care that a minimum number of qualified suppliers participate in these schemes to ensure a competitive market.

- Procurement should be only from suppliers that have complied with the WHO GMP requirements.

At the international level:

- The exchange of information and advice on successes and failures of national or pooled procurement systems, routinely updated price lists, and systems of distribution and supply will be valuable in establishing new agencies or reforming those that already exist. Bringing together data from many countries on current and anticipated needs and priorities will create a basis for producers to provide appropriate supplies. International standards for operating procurement agencies are needed, and ways to prequalify procurement agencies that attain these standards should be developed.
- The WHO, the GFATM, and the World Bank should provide leadership in meeting capacity-building demands. The WHO, the World Intellectual Property Organization, the WTO, and especially competent nongovernmental experts should provide country-level guidance on the effects of intellectual property protection on access to medicines. The goal of all technical assistance should be to strengthen national systems to be able to protect and promote public health, particularly for the poor and marginalized. Countries that do not have sufficient regulatory capacity in the short or medium term should have access to international bodies, norms, and standards to help them make efficient decisions about quality assurance, quality control, and registration.

Promoting the safety of medicines

Substandard medicines present a serious problem, especially in developing countries that have limited regulatory and enforcement capacities. The WHO estimates that as many as 200,000 of the more than 1 million deaths from malaria each year could be avoided if medicines were effective, of good quality, and used correctly. Use of substandard medicines endangers lives, wastes scarce resources, and contributes to the development of resistance to anti-infectives.

The WHO also reports that the U.S. Food and Drug Administration estimates that more than 10 percent of medicines in circulation in both developed and developing countries is counterfeit. A WHO survey of counterfeit medicines reports from 20 countries showed that 60 percent of counterfeits (products that are deliberately and fraudulently mislabeled with respect to identity

Countries can combat the sale and use of poor-quality medicines by raising public knowledge and empowering consumers

or source) were found in poor countries and 40 percent in industrialized countries (WHO 2003d). A recent report from the U.S. Pharmacopeia and Drug Quality and Information programs on the quality of anti-infectives in Asia indicates that the availability of substandard medicines (genuine products that do not conform to the pharmacopial standards set for them) and counterfeit medicines has reached a disturbing proportion in resource-poor settings (USP and DQI 2004). This report identified gaps and weaknesses:

- Weak national drug regulatory authority and weak enforcement.
- Little or no GMP compliance by manufacturers.
- Limited laboratory capacity in terms of qualified staff and equipment.
- Lack of competent medicines inspectors.
- Lack of inexpensive, quality-assured medicines (USP and DQI 2004).

At the national level:

- Countries can combat the sale and use of poor-quality medicines by raising public knowledge and empowering consumers to demand quality assurances, conducting additional inspections on companies suspected of producing or importing substandard or counterfeit medicines, strengthening drug laws, imposing stiffer penalties for offenders, increasing postmarketing surveillance, and restructuring the drug regulatory system. However, governments that rely on donor funding find themselves constrained in calling for system strengthening and increased staffing, given current conditionalities imposed as preconditions for loans or debt relief that can limit social sector spending, especially on government staffing.
- National systems that monitor suspected adverse reactions to medicines need to become more effective. They should be capable of defining the overall pattern of unwanted reactions in the population (and in particularly susceptible groups) and also cases of frank injury due to medicines. Independent medicines information centers should be supported as part of improving information exchange—nationally and across borders—on medicine quality and safety. These centers must include data on benefit-risk assessment of particular agents or products, regulatory decisions involving safety issues (such as the withdrawal of disproportionately risky medicines), and reliable information on poor-quality products and producers.
- Work should be undertaken to institute no-fault systems for redressing injury caused by medicines.
- National registration should require bioequivelancy information for both originator and generic medicines, to be provided and financed by the company seeking registration.
- Registration in most developing countries takes too long for reasons that are not always clear. Registration procedures should be simple, straightforward, and equitably applied. When possible, fast-track processes

Donor support for strengthening medicine supply systems in low-income countries should be a long-term commitment

should be available for medicines for national priority health needs, especially those prequalified by the WHO.

At the international level:

- International agencies and donors need to make safety and quality of medicines a higher priority by supporting regulatory strengthening and the timely exchange among countries, whether importing or exporting, of information relating to the safety of medicines. They should also enforce compliance with international GMP.
- Recent WHO initiatives to prequalify both individual products for high-priority diseases and the factories producing these products need to be vigorously pursued and extended.
- The WHO prequalification project should be strengthened, expanded, and made a permanent and well funded function of the WHO.
- International organizations should share information about poor-quality products and producers based on reliable and accurate data and strengthen systems for sharing information on benefit-risk assessment and regulatory decisions (such as withdrawals). International organizations should also support existing adverse event monitoring systems.
- International organizations should work to strengthen national regulatory capacity through training, capacity building, information sharing, evaluation of best practices, and sustained funding.

Increasing the affordability of medicines

The medicines supply systems in many developing countries are seriously underfinanced. It would be unrealistic to imagine that developing countries will succeed in correcting this situation on their own during the coming 15–20 years, especially in light of growing disease burdens from major epidemics. Donor support for strengthening medicine supply systems in low-income countries must be a long-term commitment. This support should be coordinated through sectorwide approaches.

Adequate and fair financing

Financing strategies should promote health system strengthening and progress toward national self-reliance over time.

At the national level:

- There must be a progressive increase in the public sector budget for essential medicines, particularly to ensure improved access for the poor; this is likely to require a shift in the allocation of government resources. Political will must exist and government allocation decisions should be made using accurate data (such as those from national pharmaceutical sector baseline surveys). Governments need to understand the importance of guaranteeing financing for procurement arrangements. User fees act as an economic barrier to healthcare for the poor. They do not

Essential medicines should be provided at no cost to the end user in developing countries

provide an adequate nor long-term solution to the problem and they should be phased out wherever they exist.

- Essential medicines, along with other essential health services, should be provided at no cost to the end user in developing countries. For the poorest countries, financing in the short to medium term must come from wealthy countries, which have repeatedly committed to spending 0.7 percent of their national GDP on official development assistance and, in most cases, have fallen short. Community financing, while a useful complement to government-financed healthcare, cannot be, in the short term, a viable option for sustainable financing of primary healthcare in low-income countries.
- Payments made to providers at all levels—importers, wholesalers, and retailers—should be commensurate with the degree of service they provide, as determined by appropriate national authorities.
- The acceptance of public or private donations of medicines should strictly follow *Guidelines for Drug Donations* (WHO 1999a).

At the international level:

- The donor community needs to accept the fact that low-income countries will need substantial additional financing to meet even the most basic primary healthcare (including medicines) packages per capita. In Uganda for example, the shortfall on medicines alone is stark ($1.20 allocated per person against an estimated need of $3.50 per person). It is cruelly cynical to suggest to poor countries that they need to make healthcare a priority for sustainable social and economic development and then not deliver the financing required. Low-income countries need long-term, sustained financial support to strengthen their health systems and procure needed medicines. In many cases, they will require debt relief.
- Health sector budgets should be privileged in programs supervised by the international financial institutions, and levels of donor assistance should be adequate to support levels of service needed to achieve the Millennium Development Goals.
- Donors should fund recurrent costs, such as salaries, for the poorest countries for the short to medium term to enable health systems to function.
- In low-income countries, loans will occasionally be justified in order to provide acute relief, but in principle, funding should be in the form of outright financial grants, preferably provided without ties. Where loans are made they should be earmarked for health systems development and not for the purchase of consumables, such as medicines. The world community can also provide valuable support in acquiring, analyzing, and disseminating comparative financial data on medicines supply and the flow of finance between and within countries. This will form a valuable basis for policies designed to ensure sufficient and equitable financing.

In developing countries, the overwhelming burden of poverty means that most essential medicines are not affordable

Financing should promote integrating medicines procurement and supply with wider health policies and systems. Sharing accurate and consistent data on producer prices, markups and profits, tariffs and taxes, and fees and other charges will promote transparency and provide a stronger base for effective analysis of actual costs and financing needs. Middle-income developing countries should be given incentives to allocate more of their available national budgets for healthcare and medicines, with some international support being an option as needed.

- Innovative new global mechanisms to promote pharmaceutical R&D for urgent health problems of the developing world should be a priority. Although the total amount required is not clear, assessing current international funding flows and existing R&D needs is urgently required to identify the magnitude of funding required.

Countering high prices

Prices matter. If a price is set at a level that a consumer cannot afford, the medicine will not be bought and used. In developing countries, the overwhelming burden of poverty means that most essential medicines are not affordable. Yet every day poor people risk their tenuous economic security to purchase medicines. Too often, the decision is a brutal tradeoff: food, housing, or education for a child or the purchase of needed medicines. Sometimes medicines are unavoidably costly, but in a great many instances they are disproportionately expensive. The reasons for high prices are multiple, and the problem therefore has to be tackled vigorously at various levels. Market competition remains the most potent way to affect and lower prices. Additionally, the presence of an effective and efficient procurement and distribution system cannot be overemphasized.

At the national level:

- Governments have a range of tools available to help manage and lower medicines prices: use available and impartial price information; have and use an updated essential medicines list; have a pro-generics approach in policy (including mandatory substitution), planning, and procurement; promote price competition in the local market; promote bulk or pooled procurement (while taking care to maintain adequate numbers of qualified suppliers to supply the market); negotiate equitable prices for patented essential medicines; eliminate taxes (such as VAT), duties, and tariffs on essential medicines where feasible; minimize markups; encourage local production of essential medicines where feasible; and ensure TRIPS public health safeguards are in national legislation and the expertise and will exist to use them.
- Prices for medicines should be transparent because information asymmetries are a main source of procurement inefficiencies that can result in higher prices. Medicines price lists, such as those published by the WHO and MSH, can be a valuable tool for countries.

Prices for medicines should be transparent because information asymmetries are a main source of procurement inefficiencies

- Prices of medicines in developing countries must be reduced to the minimum sustainable level, which in many developing countries means that industry needs to provide these medicines at production cost ("no profit, no loss") to national health systems. In middle-income countries, differential pricing should be pursued, although the prices will not be at marginal cost.
- Governments should recognize that guarantees of timely payment and financial credibility with suppliers are extremely effective for lowering prices. Suppliers, above all, want to know that they will be paid and that it will be in a timely manner.

At the international level:

- There is the need to identify and adopt strategies that will permit continued production and supply of low-cost generic medicines for poor populations after the January 2005. This is likely to involve providing new options beyond those already incorporated in the TRIPS agreement. Of key concern will be the impact of TRIPS compliance by India, a major source of low-cost generic essential medicines in developing countries, and overall use of the August 30 decision, which may prove too cumbersome to be considered a real solution (see figures 2.1 and 2.2 for details of how Least Developed Countries and developing countries can use it). Regional and bilateral trade agreements should not compromise the ability of developing countries to invoke the flexibilities provided in TRIPS (see, for example, Vivas Eugui 2003). The impact of TRIPS compliance and the August 30 decision on access to medicines in developing countries should be monitored by competent authorities, such as the WHO, and findings to date and recommendations reported by the end of 2007.
- Pharmaceutical companies should be willing to negotiate medicines prices based on a concept of equity.[1] Differential pricing negotiations should be simplified and transparency should be assured.
- All efforts must be made to continue and strengthen best price, assured quality procurement policies in the GFATM and the World Bank MAPs. Bilateral programs that restrict procurement to only originator medicines limit the impact of such aid to populations in great need, and such restrictions should be avoided.
- Both within exporting states and in international consultation, policies should favor international competition in the pharmaceutical field, including unhampered competition among individual firms and among originator companies on the one hand and among generics producers on the other.
- Much benefit will be gained by sharing information among countries and agencies on producer prices, markups and profits, tariffs and taxes, and other charges, so that successful approaches to reducing consumer prices in one country can be emulated in others.

The essential medicines list and standard treatment guidelines should be the basis for monitoring the appropriate use of medicines

Promoting the appropriate use of medicines

Medicines are not fully available to a population unless the treatment in which they are used is provided in such a way that the patient is most likely to benefit. In many situations, inappropriate prescribing, dispensing, and consumption of medicines means that this aim is not achieved.

Better prescribing and dispensing

In too many cases, prescribers write too many prescriptions, and they do so for many reasons. For example, quick and affordable testing for an acute respiratory infection may not be available, so a clinician will presume the worst and prescribe an antibiotic—just in case. Consumers often judge the quality of care by whether or not they received a prescription. In some developing countries, the average number of prescriptions per visit can exceed three. Multiple prescribing is not advisable in most cases. It can put the burden on consumers to decide which of the multiple items they can afford to purchase. Prescribing nonessential or ineffective medicines (such as cough syrups) is also a problem.

At the national level:

- A coordinated policy should be introduced to promote the appropriate use of medicines. There should be an essential medicines list, developed according to established international practice and reflecting the health needs and priorities of a given country. The essential medicines list should also be in line with evidence-based standard treatment guidelines. The standard treatment guidelines should also provide the basis for practice, as well as for teaching and evaluating health professionals about the use of medicines.
- Hospitals should set up medicines and therapeutics committees.
- The essential medicines list and standard treatment guidelines should be used as the basis for the ongoing monitoring of the manner in which medicines are used. Appropriate and ongoing development and modification of the standards should reflect current knowledge and country-specific challenges and responses. Similarly, dispensers need to be trained according to these standards and their performance needs to be monitored.
- The tasks of prescribing and dispensing should, wherever possible, be separated to avoid overprescribing because of financial incentives to the prescriber.
- It is vital to provide reliable information on medicines and their use, both during the education of professionals and on an ongoing basis during their professional careers (such as through the publication of formularies, standard treatment guidelines, and regular prescribing bulletins). The information provided by manufacturers and importers may supplement this, but measures should be taken to ensure that this

Governments should educate the public on priority health issues, including the proper use of medicines

information adheres, at a minimum, to the *WHO Ethical Criteria for Medicinal Drug Promotion* (WHO 1988).

- Patients should always be given basic information about the medications that are prescribed for them (including name, dosage, clear use instructions, and possible side effects). This approach will require sensitivity to patient population characteristics, such as accommodating different dialects and meeting the needs of largely illiterate populations.

At the international level:

- Donors and global agencies engaged in the health field need to work together to promote the appropriate use of medicines. The *WHO Ethical Criteria for Medicinal Drug Promotion* (WHO 1988) should be updated and extended to deal with newer issues, including the trend toward direct-to-consumer advertising and the increasing use of the Internet to promote medicines.
- The WHO should also ensure the worldwide sharing and dissemination of authoritative texts on the best way to treat major and epidemic conditions, so that these can form the basis for national guidelines.

Better use of medicines in the home

To ensure well-informed use of medicines in the home, long-term and incremental behavior change is needed. Education and culturally appropriate information on the use of medicines must be made available through suitable channels, with special consideration for illiterate and minority-language populations. It is often the case that even medicines that have been appropriately prescribed and dispensed are still improperly used in the home. Consumers may not have received adequate information about the treatment, and the labeling could be inadequate. They may not be able to read instructions. Consumers may be seeking a savings by stopping treatment when they feel better because they believe that the saved doses can then be available for use in future illness. A need exists to understand use in the household from economic, social, cultural and gender perspectives. Information that is primarily technical in nature could be missing the point of why inappropriate use is taking place and consequently could be of little use in changing behavior.

At the national level:

- Governments should educate the public on priority health issues, including the proper use of medicines. This general information should be supplemented by medicine-specific information, disseminated to households or patients in a culturally appropriate manner. This should not be a unilateral task for the authorities: community mobilization around issues of health and education is common. Forming alliances with community groups will be a valuable way of disseminating important information.

Two key crosscutting issues are the loss of trained health workers and gender barriers to access to medicines

- As is the case with commercial promotion to professionals, pharmaceutical advertising directed to the general public should adhere to accepted standards and be responsive to local concerns.

At the international level:

- International health organizations and NGOs should continue to develop and disseminate health literacy information related to appropriate use of medicines for use in developing countries.

Crosscutting issues

The Working Group on Access to Essential Medicines identified two key crosscutting issues:

- The persistent and often worsening loss of trained health workers is a threat to all efforts to improve health systems, including access to medicines.
- Gender is a key determinant in who gets access to medicines, why and how. The extent to which gender considerations are integrated in policies and programs affects their success.

Human resources

Needed improvements in medicines supply, distribution, prescribing, and dispensing are not going to be realized if the entire underlying issue of human resource requirements is not adequately and urgently addressed. At its simplest, many more skilled workers need to be trained, deployed, and retained in the healthcare system. However, as studies of human resource issues in developing countries show, the problems are daunting and complex. New approaches and substantial resources will be required.

At the national level:

- Healthcare workers need to be paid wages that will ensure they can work in the field of their training.
- Governments should develop programs that will both increase the sheer numbers of qualified workers and ensure improved distribution, especially to poorer and rural areas. In many indebted developing countries, social sector spending limits continue to impede the ability to be responsive to health staffing needs.
- Curricula for all healthcare workers involved in prescribing and dispensing should be progressively upgraded and continuing education provided.
- The community's own resource persons should be mobilized to participate in healthcare planning and delivery of large-scale treatment programs (such as vaccines programs).

At the international level:

- Important support can be provided for the training of professionals, using internationally tested curricula.
- Donor financing should be available to subsidize staff wages in critical need areas.

Medicines information should be gender responsive

- The brain drain of all types of health professionals from developing to developed countries is becoming a real crisis in some countries, such as Ghana, South Africa, and Zimbabwe. The international community needs to highlight the problem and reach consensus about how to reduce and manage the impact of this migration on developing countries. Possible solutions include banning actively recruiting health workers for developing countries or reimbursing training costs to the country that is losing that worker.
- International financing agencies, such as the World Bank and the GFATM, and major bilateral donors should focus on training and building capacity in a substantial number of supply chain managers and other essential health workers in developing countries.

Gender

Gender discrimination in all facets of women's and girls' lives has devastating consequences for their health and mortality. However, merely focusing on gender in isolation as a health issue will not succeed. The broader fundamental social, cultural, political, and economic interlocking roots of women's inequality in all societies must be tackled. Priority areas should include eliminating all forms of violence against women, especially sexual violence; improving economic security; removing discriminatory inheritance laws; and ensuring access to education for all girls.

At the national level:

- In health systems, policies and plans should mainstream gender considerations. This can be done only if women's participation increases and is valued in policymaking.
- Governments and agencies should collect sex- and gender-disaggregated data on access and use, which, in combination with adequate gender analysis, should inform policies, plans, and budgets.
- National essential medicines lists contain the recommended core medicines and devices for sexual and reproductive health recommended by UNFPA and the WHO (see box 1.1).
- Access to healthcare and treatment must be significantly increased for women and girls if the Goals are to be on track.
- Policymakers and planners would benefit from more research on gender-related aspects of medicines access and use by women and girls and men and boys. Medicines information should be gender responsive and made available in ways that are useful to women, who are most often the primary decisionmakers about healthcare and medicines in families.

At the international level:

- UN agencies and the GFATM should adopt policies and approaches that ensure that gender considerations are adequately integrated into all aspects of their planning, activities, and budgets.

Worldwide, the overall number of people without access to essential medicines has changed little in recent decades

Concluding remarks

Access to medicines has always been an important concern in health development policymaking and programming. But it was the WHO's call for "Health for All by 2000" in 1977 that launched what has been an ongoing effort to examine and eliminate barriers to access, especially for the poor. Both the frameworks and the expertise exist to understand, in complexity and scale, how to address all the major obstacles. However, to date, the world remains a long way from attaining equitable access within developing countries, let alone across regions.

Thirty years ago, medicines policy was a technical discourse mainly among UN agencies, ministries of health, and international experts. However, the growing AIDS pandemic has galvanized discussions about access to treatment. The United Nations, donors, recipient governments, and suppliers are being pressured by a growing global network of public interest NGOs and civil society groups that need medicines and are not able to get them. New bodies, such as the GFATM, have been founded to provide financing for national programs to tackle three of the major diseases of poverty. Existing organizations, both public and private, have become increasingly engaged in finding new ways to increase access to medicines. But more needs to be done, and it will require new thinking and new approaches.

In the last decade, most developing countries have taken measures to improve access to medicines, with varying degrees of success. Even where there have been setbacks, the experience gained strongly indicates that progress is possible. Where both the initiatives and the results have been monitored, lessons emerge that can be adapted to local conditions and applied elsewhere. A key finding is the need to involve the community in developing health system policies and programming.

Not all trends are developing satisfactorily. Finance is still seriously insufficient, and health sectors in developing countries are often severely underdeveloped, without the human and fiscal resources needed to respond to even the the most rudimentary health needs of entire populations. The dynamics that exist between the various agencies and institutions of the international community, the political and social agendas of national governments, and the needs and demands of communities and grassroots organizations remain extremely complex. In addition, the continued advance of AIDS in all of the poorest countries threatens to overwhelm already weakened, limited, and inequitable health systems.

According to the WHO, access to essential medicines worldwide increased from 2.1 billion to 3.8 billion between 1977 and 1997. A closer look at the numbers, however, shows that the overall number of those without access remains unchanged and that these people are primarily the poorest and most marginalized. Consequently, it remains to be seen if current knowledge about access and current approaches to increasing access adequately reflect a truly pro-poor framework from the global level (especially international financial institutions, UN agencies, and development strategies of major donor countries) to the local level (including real commitment by national governments to tackle poverty and take steps to improve national economic development).

A sustainable, consistent supply of medicines is both an indicator of overall health system functioning and the means by which these systems can grow and strengthen. Access for all to supplies of medicines, especially for those people who have been traditionally marginalized, is indicative of a much more profound, positive, and socially transformative process. No single institution, organization, industry, or level of government can make this happen alone. Nor can it be accomplished without the voice and partnership of the communities who bear the hardship of deaths and disability from preventable and treatable diseases. Each year, each month, without definitive and aggressive action to ensure access to quality health services, including essential medicines, for the people who live in poverty increases the magnitude of the challenge that must be overcome to begin to reverse the major epidemics. Major increases in investment, focused research efforts, open and transparent processes, and an international and national commitment to act are needed now.

Country case study

Access to medicines in Uganda: intersections with poverty

Medicines, as global commodities, are vulnerable to market forces and the legislated constraints imposed by international trade and patent regulation, as well as to country-level import/export taxes, tariffs, and regulations. Just as these larger issues are contextualized within the current global macroeconomic situation, issues of accessibility, availability, and appropriateness will also be contextualized within a constellation of culturally significant factors (such as economic status, gender roles, stigmatization of diseases, geographic location, and ethnicity). These socioeconomic features will largely determine the ability of specific individuals and families to access the medicines that are available in any given location. Therefore, ensuring access must be understood as the processes of making medicines available through regulation, importation, distribution, and safe prescribing and of identifying and removing the barriers that must be locally defined and addressed. This case study seeks to integrate some of these features within a comprehensive frame of reference, one that more closely aligns to the integrated experience of individuals and communities. This is a very brief examination of these issues, based on existing data. A detailed description and analysis for any country or specific region within a country would require a report substantially larger and more in-depth than is possible here.

In 2002, the Ugandan Ministry of Health, in partnership with Health Action International Africa and the World Health Organization, completed the *Uganda Pharmaceutical Sector Baseline Survey* (Uganda 2002b). This proved to be a valuable resource for the UN Millennium Project Working Group on Access to Essential Medicines when it met in Kampala, Uganda, during the summer of 2004. This document and the input from various other agencies and individuals during the course of the meeting and afterward provided the task force members with valuable insight into the issues of access to medicines in Uganda. We wish to thank Mr. Joseph Serutoke, Essential Drugs Advisor, Essential Drugs and Medicines Policy, WHO Country Office in Uganda, for his informative presentation during the meeting. Mr. Serutoke was also instrumental in our access to reports published in Uganda and elsewhere and he provided valuable feedback in the formulation of recommendations. Thanks are also in order to Kevin Burns of Partners In Health, who conducted extensive research for the development of this case study.

Uganda

Uganda is a landlocked, equatorial East African country that borders the Democratic Republic of Congo, Kenya, Rwanda, Sudan, and Tanzania, with Lake Victoria making up much of its southern border (map A1.1). Following a long period of, at times, violent political turmoil, Uganda developed a democratic constitution in 1995, and held presidential and parliamentary elections in 1996 and 2001. The current president is Yoweri Museveni, in office since 1986.

The United Nations estimates the population of Uganda at 24.2 million (2004) with 88 percent living in rural areas (table A1.1). Classified as a low-income country by the World Bank, Uganda is also classified as a Least Developed Country by the United Nations. Per capita income (using the Atlas method) was reported at $240 in 2003, reflecting a continuing decrease. As of 2002 Uganda's total external debt amounted to $3.8 billion. This heavy debt service is one reason the country struggles to provide social services in adequate quantity and quality (Sachs and others 2004). The economy depends largely on agriculture, with 80 percent of Ugandans deriving their livelihoods from this sector. Currently, it is estimated that 35 percent of Ugandans live on less than a dollar a day and are unable to meet their basic requirements.

Displaced populations

There continues to be isolated unrest in the northwestern region of the country where a long and violent conflict has continued for 18 years. Almost the entire population of the Acholiland region (estimated at 1.8 million people) has been displaced into poorly equipped relocation camps. Relying on international agencies for the most basic necessities, this large population of displaced people is increasingly vulnerable to sickness and disease and without reliable access to medical services or essential medicines.

Burden of disease

Like many countries in Sub-Saharan Africa, the issue of universal access to medicines has taken on a renewed urgency and visibility in recent years due to the AIDS, TB, and malaria epidemics (table A1.2). However, these three diseases are not the only ones confronting people in this or any other part of the region. "Diseases of modernity"[1] such as adult-onset diabetes, hypertension,

Map A1.1

Uganda: Population density, 2000

People per square kilometer

Source: Center for International Earth Science Information Network (CIESIN) and International Center for Tropical Agriculture (CIAT). Copyright Trustees of Columbia University, City of New York.

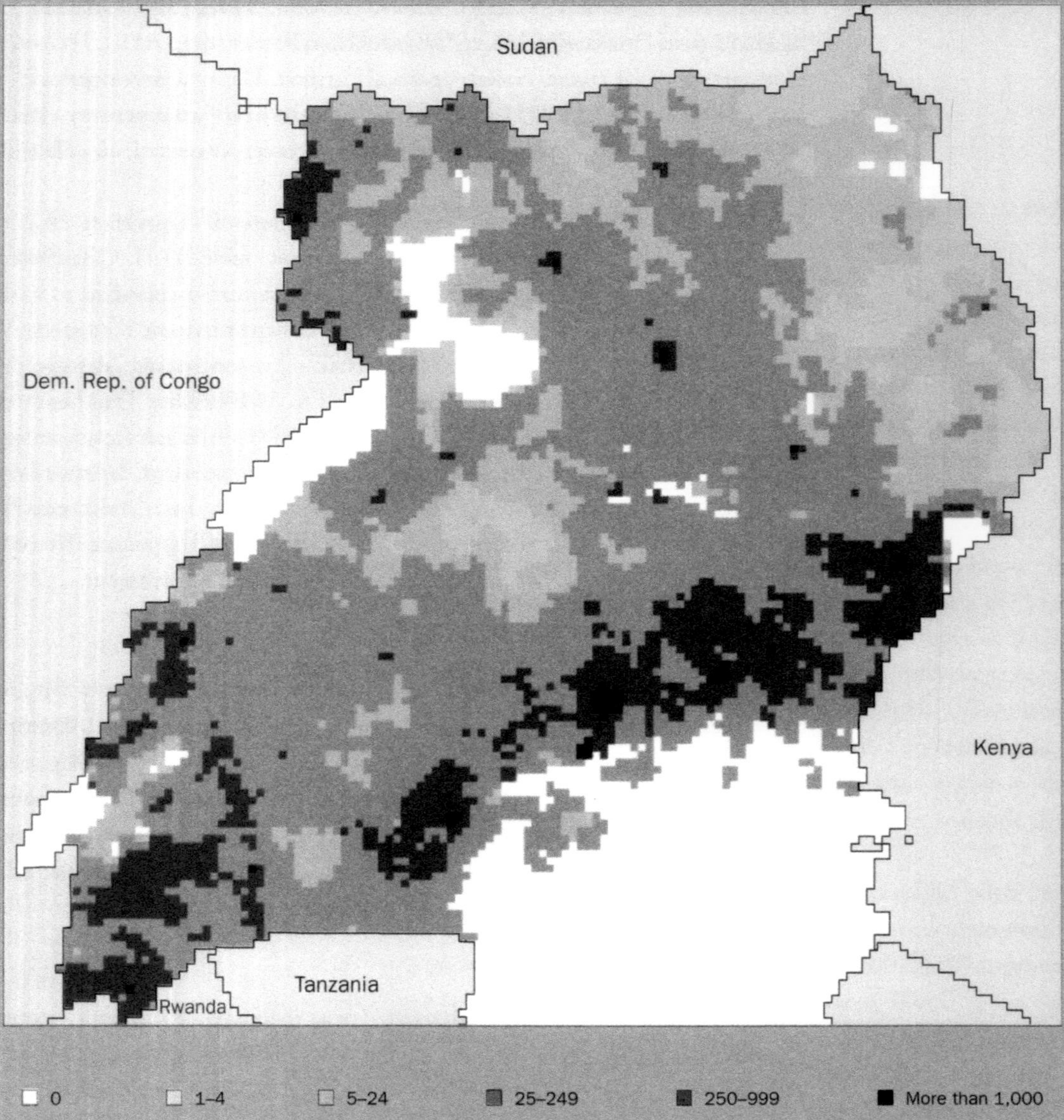

Table A1.1

Uganda: select socioeconomic indicators

Population in 2004: 24.2 million

	1999	2000	2001	2002	2003
Rural population (percent of total population)	—	87.9	—	87.8	87.3
Roads, paved (percent of total roads)	—	6.7	—	—	—
Literacy rate					
Adult female (percent women > 15 years)	—	—	—	59.2	—
Adult male (percent men > 15 years)	—	—	—	78.8	—
Youth female (percent women ages 15–24)	—	—	—	74.0	—
Youth male (percent men ages 15–24)	—	—	—	86.3	—
Primary education completion rate, girls (percent of relevant age group)	—	—	—	61.7	—
Primary education completion rate, boys (percent of relevant age group)	—	—	—	72.9	—
School enrollment, tertiary, women and girls (percent gross)	—	—	2.23	—	—
School enrollment, tertiary, men and boys (percent gross)	—	—	4.26	—	—
GDP per capita (Atlas method, current $)	—	280	—	—	240
Percent of population living below $1 per day (1990–2001)	—	—	—	—	35
Income share					
Held by lowest 20 percent	5.9	—	—	—	—
Held by highest 20 percent	49.7	—	—	—	—
Unemployment, women (percent of female labor force)	8.0	—	—	—	—
Unemployment, men (percent of male labor force)	6.7	—	—	—	—
Labor force, women (percent of total labor force)	47.6	—	—	—	47.5
Labor force, children ages 10–14 (percent of age group)	44.1	—	—	—	42.9

— Not available.

Source: World Bank 2004b.

and cardiovascular disease join an array of infectious and parasitic diseases that are almost unheard of throughout the developed world, including hemorrhagic dengue, yellow fever, filariasis, leishmaniasis, onchocerciasis, trypanosomiasis, Rift Valley fever, and schistosomiasis.[2]

> Until 1995 AIDS was the greatest health challenge facing Uganda, in addition to malaria and other diseases. In 1996 there was a substantial decline in the national HIV/AIDS prevalence rate, from 20 percent in 1991 to 6.5 percent in 2001, which made Uganda a model example internationally in combating HIV/AIDS. In this regard, Uganda moved ahead of the international target for the [Millennium Development Goal] on HIV/AIDS. Whereas it aims at halting and beginning to reverse the spread of AIDS by 2015, Uganda met this target in 1996, almost twenty years ahead of schedule. However, the important challenge is that complacency seems to have set in, which might cause a reversal in the downward trend. Recent evidence reveals that the prevalence rate increased from 6.1 percent in 2000 to 6.5 percent

Table A1.2

The burden of three major infectious diseases in Uganda

	2000	2001	2002
Total TB prevalence[a]	544	—	—
TB incidence (all cases)[a]	377	—	—
TB mortality[a]	61	—	—
HIV prevalence (15–49 years)[a]	5.8	5.4	—
Estimated AIDS-related deaths	—	84,000	—
Estimated number of people living with HIV/AIDS			
Adults	—	600,000	—
Children (0–14 years)	—	110,000	—
Percent of adult (15–49 years) TB cases that are HIV positive	—	24	—
Malaria mortality[a]	151	—	—
Clinical cases reported			
Older than 5 years	—	—	1.16 million
Younger than 5 years	—	—	720,298

— Not available.

a. Reported per 100,000 population.

Source: WHO 2004c.

in 2001. Although awareness of HIV/AIDS is widespread, knowledge of ways of avoiding the virus are not as well known. According to the 2000 Uganda Demographic and Health Survey, 13.4 percent of Ugandans did not know any programmatically important way to avoid HIV/AIDS (Sachs and others 2004, p. 188).

In a 2004 report on Uganda (WHO 2004c), the WHO cited the 1995 *Burden of Disease Study* and *The Uganda Demographic and Health Survey* (Uganda 2001), which showed that 75 percent of years lost to premature death are due to 10 preventable diseases. More than 60 percent of the total death burden was attributed to perinatal and maternal conditions (20.4 percent), malaria (15.4 percent), acute lower respiratory tract infections (10.5 percent), AIDS (9.1 percent), and diarrhea (8.4 percent). Others include tuberculosis, malnutrition, anemia, intestinal worms, trauma/accidents, skin infections, and dental health.

Poverty in Uganda

Poverty remains Uganda's major development challenge. The growth the country recorded in the 1990s initially led to a reduction in poverty from 56 percent in 1997 to 35 percent in 2000. However, by 2002 poverty had increased to about 38 percent (Sachs and others 2004). Uganda's government has been in the forefront of developing policies to reduce poverty. It was one of the first countries to prepare a comprehensive strategy for poverty reduction using a participatory approach. During 1995–97 Uganda developed the Poverty Eradication Action Plan to create a framework for poverty eradication by 2017. The revised plan (2000) was presented as its Poverty Reduction Strategy Paper (PRSP). Among other reasons, this was done to comply with the eligibility requirements for future debt relief under the Highly Indebted Poor Countries

(HIPC) II Initiative of the World Bank and IMF. Within this framework, the government of Uganda has developed a review process every three years as well as focused strategies to address poverty reduction in every sector of the government. The Poverty Eradication Action Plan has been used as a major guide to allocation of resources within the government (Uganda 2002a).

Part of the review process includes Participatory Poverty Assessments in which the members of 12 districts (out of the 56 districts in the country) report on the progress and impact of the government's poverty reduction strategies on their lives, families, and communities. These assessments bring together representatives from the Ministry of Finance, Planning, and Economic Development, district authorities, civil society organizations (both NGOs and academic institutions), and donors. In the Second Participatory Poverty Assessment Process (Republic of Uganda 2002), the most frequently cited cause of poverty was poor health and diseases. Improving access to essential medicines within the context of expanded and improved health services will be a key component in any effort to reduce poverty and inequality (table A1.3).

Health sector

Resources (both human and financial) as well as access are two major challenges within the health system. It is estimated that at least 53 percent of the population lives less than 5 kilometers from a health facility, with a range from 9 percent in parts of Kitgum (near the northern border with Sudan) to 100 percent in Kampala. In 2002 42 percent of approved posts in the public health sector were filled by trained workers, up from 33 percent in 1999. The human resource needs in Uganda remain significant and long-term planning for scaling up nursing and medical education is a key component of Millennium Development Goal planning.

There are only about 270 qualified dispensers or pharmacy technicians in Uganda. The current educational system produces, on average, 15 a year. At this rate, it will take Uganda over 50 years to achieve a ratio of one pharmacy technician for every 20,000 people. The situation with registered pharmacists is even worse: with an output of 10 per year, and with only 171 registered pharmacists in Uganda, it will take the government over 90 years to reach a ratio of one pharmacist for every 20,000 Ugandans (Kibumba 2002).

President Museveni initiated a major policy change in March 2001 when he abolished user fees in government health units. The public response was immediate and resulted in a dramatic change in the health system, with poor people benefiting disproportionately—the lowest quintile captured 50 percent of the benefits (Deininger and Mpuga 2004). Currently, other countries are considering this policy. The effects on direct health services and outcomes became apparent almost immediately and included healthcare reforms that were introduced and accelerated to meet the new demand. In addition to illustrating the restrictive impact of user fees, the increase in use has highlighted infrastructure

Table A1.3
Select health-related indicators in Uganda

	2000	2001	2002
Annual growth rate (percent)	—	—	3.0
Life expectancy, female (years)	—	—	50.8
Life expectancy, male (years)	—	—	47.9
Infant mortality rate (under 1 year)	—	—	82
Maternal mortality ratio per 100,000 live births	880	—	—
Births attended by skilled health staff (percent of total)	—	39	—
Malnutrition prevalence, weight-for-age (children under 5 years)	23	—	—
Immunization DPT (percent of children 12–23 months)	53	—	72
Immunization measles (percent of children 12–23 months)	56	—	77
Health expenditure per capita (US$)	14	14	—
Health expenditure			
Public (percent of GDP)	3.1	3.4	
Private (percent of GDP)	2.5	2.5	
Total (percent of GDP)	5.6	5.9	
Improved sanitation (percent of population with access)	79	—	—
Rural (percent of rural population with access)	77	—	—
Urban (percent of urban population with access)	93	—	—
Improved water source (percent of population with access)	52	—	—
Rural (percent of rural population with access)	47	—	—
Urban (percent of urban population with access)	80	—	—

— Not available.

Source: World Bank 2004b.

problems as health facilities struggle to meet the increased demand for services. The government of Uganda recently found that lack of medicines is now identifed as the most serious problem in health units (Republic of Uganda 2002).

There are 56 districts with an administrative structure for healthcare consisting of five levels, encompassing health centers I–IV (village, parish, subcounty, and county) and the district/general hospital. There are 11 referral hospitals (which also act as district hospitals in the areas where they are located) and 2 national referral hospitals (Mulago and Butabika). Mulago and Mbara Hospitals also act as university teaching hospitals (Uganda 2002b). The Ministry of Health reports that there are 104 hospitals (57 government, 44 NGO, 3 private), 250 health centers (179 government, 68 NGO, and 3 private), 2 palliative care centers (1 government, 1 NGO), and 1,382 other healthcare facilities (989 government, 352 NGO, and 41 private).[3]

The Poverty Eradication Action Plan of 1997 was followed by the National Health Policy (NHP) in 1999, and the Health Sector Strategic Plan (HSSP

2000/01–2004/05) in 2000. Both the NHP and the HSSP focus on the Uganda National Minimum Healthcare Package, which is the basic package of services goal to which all Ugandans have access. The revised National Drug Policy (2001) aims to contribute to the attainment of a good standard of health, through ensuring the consistent availability, accessibility, and affordability of essential medicines of appropriate quality, safety, and efficacy, and by promoting their rational use.

Access to medicines

A five-year National Pharmaceutical Sector Strategic Plan for fiscal 2003–07 has been developed. The overall per capita minimum expenditure for basic healthcare provision is estimated to be $28 per person. Current spending is estimated to be a small fraction of this. Funding for medicines in 2002/03 was $1.20 per capita, which is only one-third of the estimated $3.50 per capita needed (excluding the pentavalent vaccine that is currently donated and antiretrovirals). The midterm review concludes that this shortfall poses a serious threat to sustained availability of essential medicines and health supplies, and hence to the delivery of the Uganda National Minimum Healthcare Package (Caines and others 2003, p. 10).

The National Medical Stores is the government agency charged with procuring, storing, and distributing essential medicines and supplies to the public sector. The missionary hospitals get their medicines and health supplies from Joint Medical Stores. Only when medicines and equipment are out of stock from the National Medical Stores can public health facilities obtain them from elsewhere, including the Joint Medical Stores.

Currently there are five large-scale and five small-scale pharmaceutical manufacturers. There are 2,939 public sector health facilities from which medicines may be dispensed, 215 private pharmacies, and 2,600 drug shops. Of the private pharmacies, nearly 80 percent are in the three major towns of Kampala, Jinja, and Mbarara (Uganda 2002b).

More than 90 percent of pharmaceuticals are imported into Uganda, with less than 10 percent produced locally (UNIDO–AAITPC 2001). Uganda's imports of medicines in 1999 amounted to $73,776,000; domestic production of medicines was valued at $7,440,632. Most pharmaceutical and health products are imported by the National Medical Stores (UNIDO-AAITPC 2001).

As of 1999 the WHO estimated that in Uganda 50–79 percent of the population had sustainable access to affordable medicines.[4] The Ugandan Pharmaceutical Sector Baseline Survey found that only 47 percent of surveyed public health facilities had more than 75 percent of key medicines available. One out of three facilities storing medicines were found to have "not adequate" storage, leading to medicines of "poor or doubtful quality." Given that at least 53 percent of the population lives more than 5 kilometers from a health facility and that 35 percent of Ugandans live on less than a dollar a day, the actual percentage with access to essential medicines remains to be clearly defined. In the second Participatory Poverty Assessment, medicines shortages were reported at all sites.

Importation issues

Strategies to enhance the ability of the Ugandan government to procure medicines at the best price could include differential pricing (adapting prices to the purchasing power of governments and households), bulk purchasing, competition, and skillful negotiation (WHO and WTO 2001). Mechanisms for differential pricing are actively supported by the WHO and include voluntary negotiated agreements with companies, voluntary licensing with multiple producers ("licensed competition"), compulsory licensing, and patent waivers (Quick 2003). The Commission on Macroeconomics and Health strongly supports "differential pricing in low-income markets as the operational norm, not the exception" (CMH 2001).

Another option available to Uganda as a feature of its status as a Least Developed Country is to invoke TRIPS flexibilities and the suspension of pharmaceutical patents as described in the Doha Declaration on TRIPS and Public Health (WTO 2001). The Doha Declaration, issued at the November 2001 WTO ministerial meeting, supports the right of member states to implement the TRIPS agreement in a manner that promotes public health and access to medicines for all. To fully take advantage of differential pricing, Uganda must retain its ability to issue compulsory licenses to a company in another country (CMH 2001).

After 2005, all member countries of the WTO except Least Developed Countries are required to put into force a patent system that includes both product and process patents. A 2001 review summarizes three studies that predict pharmaceutical price increases of 200 percent or more with full implementation of TRIPS requirements in developing countries (Scherer and Watel 2001). A 2002 report by the United Kingdom's Commission on Intellectual Property Rights also echoes this concern (UK Commission 2002).

In Uganda, new patent law legislation has been drafted that would make it harder to access generic medications—a key component of differential pricing. This legislation includes legal provisions that criminalize patent infringement, granting data exclusivity to prevent the registration of generic versions of a medication for a specified length of time, linking patent status with drug regulatory authority approval, and granting patent protection for new uses of previously patented products. The proposed legislation goes beyond what is required by the TRIPS agreement.

Full differential pricing, including access to generic medicines, would have a huge impact on the amount of medicines Uganda will be able to procure. An estimated 100,000 of the 530,000 Ugandans living with HIV need antiretroviral medicines. Minister of Health Jim Muhwezi recently announced that government negotiations and generic antiretrovirals have helped to reduce the treatment costs from $1,500 per person per month to $30 per person per month. This has resulted in an immediate increase in antiretroviral access to almost 20,000 people, with a projection of 60,000 people having access to these medicines on the arrival of projected donor funds.[5]

In addition to inadequate fiscal resources for procuring essential medicines and health supplies—currently it is estimated that only a third of what is required is being made available—institutional and human resource capacity to manage pharmaceutical supplies and service provision are inadequate. The issue of utilization, defined by one Ugandan expert as "poly-pharmacy and overuse of antibiotics and injections as well as inappropriate self-medication" is also a problem.

The view from the community level

The following comments were offered from community members during the Participatory Poverty Assessment Process completed in 2002. Full reports from the 60 communities that participated are available online.[6]

> They revealed that most of them had resorted to local herbs for treatment but they also revealed that the herbs had become very scarce and the situation was worse during dry season when most of the plants were burnt down. The study also revealed that some people went to traditional healers/witch doctors. However, it had been noted that often their powers did not work. The community resorted to these due to shortage of drugs in the government health centres and lack of money to go to private health clinics.
>
> District of Kitgum, Northern Region
> Area: 16,136 sq. km
> 2002 population: 286,122 (1.4 percent urban)

> [T]he people said they still go to the government health units despite the rudeness and drug shortages. They explained that government units had equipment and facilities that were lacking in private clinics. Also, drug shortages were reported in the government health units to the extent that people said they were sent to private clinics to purchase them. They therefore go to government units for diagnosis and prescription then go for treatment in the private health units.
>
> District of Masindi, Western Region
> Area: 8,458 sq. km
> 2002 population: 469,796 (0.8 percent urban)

> The people of Nakapelimen claimed that they took to herbs as first option for treatment because they often did not get treatment in the hospital on the account that there were no drugs most times. They

also argued that they were sometimes referred to private clinics to buy drugs and yet they do not have the cash to do that. . ..

On decisionmaking, the women say that they are not free to make independent decisions about most things—including matters of healthcare. That in spite of the fact that it was they (the women) who were responsible for treatment of the children, they had to seek for permission from their husbands to visit a health center. They said they did this in spite of most men not providing financial support for treatment. However the women of Lokileth and Naoi (villages where the nearest health facilities were NGO units) explained that they did this so that the men could sell a goat or sheep to raise some money required for payment for treatment. In Lorukumo the women explained that even when they informed them, the men they did not give any money—since even the survival of the household depend on them (the women). They explained further that since the children traditionally belong to the man, they could not take the child away from home without informing the man.

District of Moroto, North Region
Area: 14,113 sq. km
2002 population: 170,506 (0.2 percent urban)

Recommendations

The Uganda Pharmaceutical Sector Baseline Survey of 2002 includes an impressive number of specific recommendations. The following are included here because, as with many of the other recommendations, they specifically address issues of access on the community level.

1. Develop and implement strategies that ensure equitable access, affordability, and sustainable financing for health services in general and access to essential medicines in particular.

- Advocate for increased funding to the health sector in general and for medicines in particular from the government and from focused donor support.
- Strengthen institutional and human resource capacity for coordination and implementation of the National Drug Policy—setting up a functional and adequately staffed Department for Pharmaceutical Services within the Ministry of Health and similar structures at the local government levels and in hospitals.

2. Develop and implement interventions to address the poor availability of drugs in public health facilities.

- Strengthen institutional and human resource capacity to manage pharmaceutical supplies and service provision within the public health facilities.
- Establish a Logistics Management Information System and a Medicines and Health Supplies Tracking System to monitor drug utilization, facilitate accurate quantification, and harmonize procurement.

3. Identify and develop interventions to address the significant increase in the number of antibiotics prescribed per patient contact.

4. Identify, develop, and continue implementing interventions to further decrease the number of patients receiving injections.

For both (3) and (4):

- Support the formation of drugs and therapeutics committees in health facilities to coordinate the selection, procurement, and utilization of medicines at these levels, and for support for lower levels where the capacity may be limited.
- Disseminate the Uganda Clinical Guidelines to all healthcare workers in the country and train them in their correct use.
- Preservice training in health training institutions and continuing in-service medical training (may be as a precondition for licensure/practicing renewal).
- Strengthened supervision, audit, and regulation of the activities of all actors involved in the use of medicines.

5. Design and implement consumer-targeted and community-based information, education, and communication campaigns to improve rational use of medicines in the community.

- Ensure that over-the-counter medicines are dispensed with adequate labeling and provide written or oral instructions that are accurate and easily understood by laypersons. The information should include the medicine name, indications, contraindications, dosages, drug interactions, and warnings concerning unsafe use or storage.
- Conduct targeted public education campaigns that take into account cultural beliefs, illiteracy, language differences, and the influence of other social factors. Education about the use of medicines may be introduced into the health education component of school curriculums or into adult education programs, through the use of literacy courses and drama educational techniques.
- Monitor and regulate advertising, which may adversely influence consumers as well as prescribers, and which may occur through television, radio, newspapers, and the Internet.

Conclusion

Access to a sustained supply of good-quality medicines is a critical part of addressing both the current infectious disease epidemics and the long-term quality of life and productivity of people throughout the world. As an essential component of good health, it is also a human right. Reaching the people with the least access to resources and who are conditioned by the conditions of poverty to be disproportionately vulnerable to a wide array of diseases and chronic illnesses is the challenge before us. The government of Uganda continues to move forward in its determination to address these difficult issues. The international community can do no less.

Statement of dissent by representatives of the research-based pharmaceutical industry

December 6, 2004
Overview

The Millennium Development Goals (MDG) are an important symbol of global commitment to addressing the root causes of poverty. Health is a unifying theme for the Goals because progress in basic indicators of mortality and morbidity are a prerequisite for the social advances and higher economic output that, when combined, yield better life opportunities for the poor.

As we move closer toward the 2015 target date for realization of the MDGs, the R&D pharmaceutical industry is ready to demonstrate its contribution, particularly in those areas—medicines research, clinical training, and drug delivery and distribution—where we possess unique expertise. The industry is proud of its record.

We regret that the sum total of our efforts in building practical, field-based partnerships—partnerships that work—was not recognized in the final report of the Working Group on Access to Essential Medicines. Representatives from our industry actively participated in the working group from its launch until final editing was completed last month. We endorse many of the basic messages contained in the text—in fact, we contributed much of this content in cooperation with other members. Assessment of critical needs around human resources, the importance of mobilizing political will, the slow pace of donor support, filling the enormous gaps in physical plant capacity, and resolving social biases of gender and stigma are all critical factors that must be addressed in any national strategy to increase medicines access.

We believe this because we are doing it—with our own resources, ranging from the simple transfer of people-to-people skills to technical support, supply chain management, research expertise, philanthropic grants, cash, and donated goods and services.

So the question arises: after a year of effort and a record of contribution offered in good faith, why is it that we cannot sign the report?

Let it be clear that our failure to sign is not because all our views were rejected—disagreement is part of normal life—but rather because of an enormous visionary gap between ourselves and the working group in identifying root causes of the access challenge. We do not believe that the main problem in barring medicines to the poor is patent protection, nor do we accept that individual company pricing practices are fundamental to explaining why one-third of the world's poor lack access to basic, low-cost essential medicines. An inaccurate and subjective link is forged between rights, "monopoly" pricing, and global inequities in access to medicines. Much of the text on these issues was adopted from the Task Force on Trade—a group in which industry was not represented—and inclusion of this material was never discussed with the full membership of the Working Group on Access to Essential Medicines. Hence the necessity of including a detailed rebuttal to correct what we contend are critical factual errors and underlying biases on intellectual property.

We also believe that our private sector research model is worthy of preserving rather than abandoning on the risky premise that more public investment will by itself yield miracle cures against the complex scientific challenge of fighting resistant strains of infectious disease. There is a failure to acknowledge current efforts in development of drugs and vaccines for diseases primarily limited to developing countries and an unwillingness to realistically identify research gaps and establish priorities.

Most important, we contend that the skeptical stance the report takes toward the industry's partnership efforts are simply counterproductive, especially when examined against the working group's own mandate: to seek access solutions that work, in cooperation with the pharmaceutical industry. The working group seems not to have recognized that today our industry is managing programs that include the single largest antiretroviral treatment program in Sub-Saharan Africa as well as the region's first clinical education and training facility focused specifically on finding innovative local solutions to the AIDS pandemic.

In short, the report fails to provide the balanced and accurate perspective necessary to stimulate fresh policy approaches that could make a real difference in the lives of the poor. To allow these inaccuracies and misrepresentations to become accepted as truth and as the basis for moving policies forward does no one any service, least of all the patients who rely most on the commitments we have made. It would significantly diminish our ability to fulfill commitments to current and future partnerships and—most importantly—our capacity to produce new drugs, diagnostics, and vaccines.

Thus we have no choice but to respectfully dissent from the working group's report and to summarize below those four areas of greatest concern.

Specific issues and areas of dissent

Role of research-based pharmaceutical companies and the importance of

public-private partnerships in access to essential medicines in developing countries

The report does not adequately reflect the true breadth of what the R&D industry actually does to fight disease and promote the health and wellness that aids development. It is critical that this be understood because much of what is recommended in the report regarding intellectual property, pricing, and innovation would, if implemented, threaten the industry's ability to maintain existing and future contributions.

The most significant contribution research-based pharmaceutical companies can make to the goal of expanding access to essential medicines is to discover and develop new medicines. Private research-based pharmaceutical companies have produced the overwhelming majority of new medicines now on the market and available to patients in all countries. Fully 77 percent of the products approved for the WHO Essential Drugs List between 1977 and 2002 originated or were substantially developed by the R&D industry. Since 1988, 22 separate HIV-related medicines have been developed by research-based pharmaceutical companies and in 2003 there were still 87 in development. Because of the rapidity with which HIV is developing resistance to existing drugs, the need and urgency for new drugs is greater than ever before.

We are heavily invested in research and development, spending approximately 17 percent or more of sales on R&D—three times more than the next high-spending industry (telecommunications), four times more than the defense industry, and four times more than all other industries (*Pharmaceutical Industry Profile 2003*, www.phrma.org). In 2003 the combined investment in biomedical research in the United States by both the public and private sectors exceeded $70 billion, of which half ($35 billion) originated in the pharmaceutical industry.

What is generally not understood is the essential, unique, and complementary roles that both industry-funded (private sector) and federally funded (public sector) research play in translating these advances into tangible new treatments and the interrelatedness and synergies between the two. Although scientists in research-based pharmaceutical and biotechnology companies contribute significantly to basic research and thus to increasing our fundamental understanding of disease, it is also true that federally funded investigators have traditionally conducted the bulk of basic biological research. However, the pharmaceutical industry continues to lead the way in the applied research activity that ultimately results in the discovery and development of most new medicines, i.e. the actual compound or biological entity that is the drug.

Increasingly, research-based companies are valuable partners with researchers from academia and the public sector on basic research projects. However, companies that license inventions from universities still pay the majority of the innovation's final cost and pay for all the failed efforts and blind alleys (there is a 90 percent failure rate from target identification to product launch) and the promising drugs that prove not to be sufficiently safe and sufficiently effective to gain approval for marketing from the regulatory authorities (there is a 50 percent failure rate in Phase III) (*Pharmaceutical Industry Profile 2003*, www.phrma.org; Economist.com 4/21/03; Edwards, M.G., F. Murray, and R. Yu, *Nature Biotech* 2003(21): 618–25).

At the global level, the relationship between governments, multilateral institutions, NGOs, and the research-based industry has developed progressively. The pharmaceutical R&D industry plays a far larger role in the advancement of global health than is generally realized. The estimate for pharmaceutical companies' total foreign assistance for 2003 is just over $2 billion. This estimate is based on the dollar value of their product donations and cash contributions for global health programs. In 2003, Partnership for Quality Medical Donation reports that its members provided more than $1.4 billion in donated drugs, and in 2002 pharmaceutical companies' select humanitarian programs totaled $810 million (*Improving Health in the Developing World*, PhRMA publication, January 2004).

When compared with the annual budgets of governments and international health organizations, total aid from the industry is actually the same or even higher (e.g., WHO's annual program budget of $421.3 million; or UNICEF's annual budget of $1.3 billion) [see "The Privatization of Foreign Aid", *Foreign Affairs*, November/December 2003, Adelman; "The Full Measure of Foreign Aid", "Foreign Aid in the National Interest: Promoting Freedom, Security, and Opportunity", Chapter 6, Adelman, USAID/W, 2002].

More recently, the industry has worked with the WHO and other stakeholders to lower the price of key medicines like antiretrovirals and those for treating MDR-TB, and has expanded donations and technical assistance programs to help fill the capacity gap. A few examples include the African Comprehensive HIV/AIDS Partnership (ACHAP) in Botswana, a collaboration of the government of Botswana, the Bill & Melinda Gates Foundation, and Merck & Co., Inc. (26,000 patients are now receiving antiretroviral therapy through ACHAP); Pfizer built Uganda's first Infectious Disease Institute in Kampala to provide a training site for physicians and technicians drawn from throughout Africa, thus seeding best practices regionwide; Bristol Myers Squibb has initiated HIV/AIDS programs in 5 South American countries which provide antiretroviral medicines, as well as inputs into health management, medical research and education, community education and outreach, and capacity-building programs for women and children.

Likewise, the UN/Industry Accelerating Access Initiative (AAI) is a cooperative endeavor of seven research-based pharmaceutical companies, UNAIDS, WHO, UNICEF, the UN Population Fund, and the World Bank. In July 2004, WHO estimated that 440,000 AIDS victims were receiving antiretrovirals in the developing world, including 136,000 in Brazil alone. Outside of Brazil nearly 50 percent of the remaining patients were obtaining their AIDS drugs through the efforts of the Accelerating Access Initiative program. And

Boehringer Ingelheim, Merck, and GlaxoSmithKline, among others, have extended voluntary licenses to generic manufacturers in South Africa so that antiretrovirals can be manufactured for Africa—an example of the kind of "patent flexibility" that the report cites as entirely lacking from the industry.

In a program to address MDR-TB, Eli Lilly and Company is transferring manufacturing technology and technical assistance to manufacturing companies in India, China, and South Africa. In addition to supporting a center of excellence for training in Russia and establishing a surveillance program in partnership with the Centers for Disease Control and Prevention and WHO, Lilly is also providing at significantly below cost two of the five antibiotics used for treating MDR-TB to the WHO.

In contrast to popular belief, the research-based pharmaceutical companies continue to play an important role in development of medicines for malaria, TB, and other diseases occuring primarily in developing countries. For example, Novartis recently established a research center for drug discovery and development for tropical diseases in Singapore, focusing initially on TB and dengue fever. When drugs are finally produced, they will be sold at no profit. Astra Zeneca has created a new discovery research facility in Bangalore, India, which will focus exclusively on TB. GlaxoSmithKline has a dedicated facility in Tres Cantos, Spain, for drug discovery in diseases of the developing world, including malaria and TB. They currently have two antimalaria drugs in development (phases I and III) and also have vaccines in clinical trials for TB and malaria. Much of this work is in association with public-private partnerships that offer a new and innovative approach to drug and vaccine development—a development barely cited in the report.

Priorities for innovation

The report neglects the important point that most diseases that disproportionately affect low-income countries can be treated or prevented with easily available existing resources, such as medicines from the WHO Essential Drugs List, of which over 95 percent are off-patent. Yet these diseases continue imposing important burdens on the health of millions of people affected, proving that the biggest challenge that remains is getting the interventions to the people who need them. Any considerations regarding the need for novel interventions and setting research priorities should therefore be made within this specific context (IFPMA 2004).

One of the most pressing needs is to set a clear and irrefutable priority around access to medicines to fight the three biggest killer diseases: HIV/AIDS, TB, and malaria. Millions die each year from these diseases and millions more contract them in even more deadly combinations. Both trend lines and absolute numbers for these three diseases continue to rise at alarming rates among the world's population, and the 2 billion poorest people in the developing world experience the most direct and deadly impact.

Due to the speed at which these diseases are spreading and antimicrobial resistance is developing, failure here could render access to other medicines and related issues a moot point.

Partnerships built around public and R&D industry engagement are rising to this need. New and emerging infections have triggered substantial investment in research in infectious diseases in both rich and poor countries over the last decade. For example, the budget of the National Institute of Allergy and Infectious Diseases of the National Institutes of Health in the United States has increased from a little more than $1 billion in 1998 to more than $3.5 billion in 2003 (NIAID, NIH 2004). Sixty percent of this budget is dedicated to infectious and parasitic diseases relevant to developing countries (WHO 2004e). An increasing amount of their budget is to promote collaborations with scientists in other countries and to establish research centers in those countries, particularly as it relates to HIV, TB, malaria, and other parasitic diseases. The 10/90 Report on Health Research 2003–2004 (www.globalforumhealth.org) and the workshop of the Initiative on Public-Private Partnerships for Health, "Combating Diseases Associated with Poverty: Financing Strategies for Product Development and the Potential Role of Public-Private Partnerships," indicate that substantial progress has been made in the last decade by the establishment of a new type of public-private partnership for product development, as already discussed.

A small number of diseases of the poor still need R&D investment as no effective and safe treatments for them exist. For others, medicines exist but capacity-building approaches and strategies require re-evaluation. Various attempts to prioritize the need for R&D among "neglected" diseases have led to a unanimous conclusion that efforts should be focused on the three kinetoplastid diseases: African trypanosomiasis, Chagas disease, and leishmaniasis.

Surprisingly, the working group's report does not acknowledge this, nor does it accept that important progress has been made in both drug development and increases in public funding of basic research for these and other diseases primarily occuring in developing countries. A proper perspective is critical: so called neglected diseases often do not represent the most pressing public health priorities in low-income countries. In fact, they constitute a small fraction of the total disease burden. According to the 2002 *World Health Report,* tropical diseases accounted for only 0.5 percent of deaths in high-mortality poor countries, and only 0.3 percent of deaths in low-mortality poor countries (WHO 2002g).

Even for the truly neglected diseases, progress is being made through public-private partnerships and independent efforts. For African trypanosomiasis an initiative between WHO and three pharmaceutical companies—Aventis, Bayer, and Bristol Myers Squibb—has been established. There are several products available for leishmaniasis developed by pharmaceutical companies working with the WHO Tropical Disease Research group. For Chagas disease, Roche has donated rights and technology to manufacture benzonidazole (the most effective drug for Chagas disease) to the Brazilian government.

With respect to other tropical diseases, schistosomiasis can be treated with praziquantel at a cost of 30 cents per child per year. Onchocerciasis is controllable with ivermectin, and a range of treatments is available for lymphatic filariasis. The only significant tropical disease for which there is no existing medicine is dengue fever, but even for this disease there are five compounds currently at a state of discovery and preclinical development, a further two in Phase I trials, and one more in Phase II trials (IFPMA 2004).

To conclude, there is a continuing need for new and innovative medicines and vaccines to keep pace with current and emerging health challenges. All agree that the existing healthcare inequities of the developing world require new thinking. Unfortunately, quick solutions are usually equally shortsighted and fail to understand or address the complexity of the issues, and incur disastrous and potentially irreversible long-term consequences. It requires the good-faith effort, intelligence, and commitment of all parties to the development of solid, sustainable, win-win policies from which short- and long-term strategies can be developed for intervention and relief.

Inaccuracies and biases on intellectual property and pricing

Scaled-up efforts by the public sector to support developing countries; donations or voluntary differential pricing schemes by pharmaceutical companies, when combined with safeguards against diversion; and creative use of intellectual property to promote public-private partnerships around drug discovery are all critically important to building access. The report of the working group discounts or rejects all but the first of these, but the reality is that all are interrelated—without the spur to development provided by property rights and the rule of law, donor efforts to fund medicines are likely to yield only short-term gains in health.

There is now a track record on how intellectual property rights may be properly managed to support access and economic and social development. Certainly not all questions have been answered as to how intellectual property rights should be managed in connection with procurement and all medicine development programs. None of these matters is susceptible to a "cookbook" or "one-size-fits-all" approach.

We do believe, however, that the time has arrived to put behind us whether intellectual property has a productive role to play in access and to move on to the task of making it work to play such a role (Wilder and Solovy 2004; Roy Widdus, *Product Development Partnerships on 'Neglected Diseases': How They Handle Intellectual Property and How This May Contribute to Improving Access to Pharmaceuticals for HIV, TB and Malaria*, www.iprsonline.org/unctadictsd/bellagio/docs/Widdus_Bellagio3.pdf).

Throughout the report there is a fundamental bias that intellectual property is a problem to be overcome rather than a tool to be managed and used to accomplish desired goals. This bias in the report is found in recommendations to focus on the exceptions to intellectual property rights as an end in itself, rather than a necessary adjunct to a fair and functional intellectual property system. In short, the report gives far more attention and credence to the exception to the rule of intellectual property than it does to the rule itself.

This discrepancy is highlighted in the section of the report on "overarching barriers" to access to affordable new medicines and vaccines. This section was apparently added to the final report of the Working Group on Access to Essential Medicines based on discussions that took place in the Task Force on Trade.

Specifically, there is a factual reference that is not fully understandable in context—that is, "after January 2005, generic production in India and China, the source of many vital existing medicines for developing countries without productive capabilities, will be subject to TRIPS provisions." China made substantial reforms to its patent system in 1993, including providing patent protection for pharmaceutical products, and as a new member of the WTO, its intellectual property laws had to be fully TRIPS compliant as of its WTO accession in December 2002.

This is in contrast to the situation in India, which was given until 2005 to fully comply with TRIPS. China has had a patent law that protected pharmaceutical products for over a decade, and has had a TRIPS-compliant law since 2002. Hence it is not clear what changes will be made after January 2005 in China. This problem regarding the perceived effect of TRIPS implementation on China appears in several sections in the report, including the recommendation on unaffordable prices.

The working group also states that the agreement on the waiver to obligations in Article 31(f) of the TRIPs Agreement "will be too cumbersome for developing countries to exploit." Making such a statement is at least premature, as WTO members are only now passing legislation to implement the waiver, let alone deal with specific cases. Further, the waiver was a carefully negotiated, drafted, and unanimously agreed-upon compromise among all WTO members.

Lastly, the "major recommendation" to mandate the WHO to monitor TRIPS compliance is in our view unnecessary. The WHO role in trade and, in particular, on intellectual property matters, has been fully discussed in the World Health Assembly over the past several years. To the extent that the WHO has a role in this area, it is already being addressed and it is clear that most member countries of the WTO are reluctant to see the WHO expand its purview into this area without proper consultation and negotiation concerning the respective areas of engagement and enforcement.

With regard to industry pricing, the report states that "governments ensure TRIPS public health safeguards are in national legislation and have the expertise and will to use them." There is no doubt that having expertise and the will to implement provisions regarding the protection of intellectual property protection are important. Indeed, for WTO members, implementing the TRIPS agreement is both required and good policy. That said, urging the implementation of "safeguards" or "flexibilities" is narrow and insufficient. Rather, the goal should be to fully and fairly implement the TRIPS agreement as a whole—including its substantive obligations and limitations and exceptions—to achieve the legislative intent of the entire agreement.

This prejudice is evident in the section on addressing issues at the international level. In particular, there is a statement of a need to provide "new options [pertaining to exceptions and limitations to intellectual property protection], beyond those already incorporated in the TRIPS agreement." Thus, rather than urging a full and fair implementation of the TRIPS agreement, the goal is to seek further ways to undermine intellectual property protection before it is fully established in many countries.

In this same section there is a statement that competition in the pharmaceutical field must be favored—"including unhampered competition between individual firms and between innovative companies on the one hand and generic producers on the other." In context, it is clear that this statement calls for an elimination of intellectual property protection. That is, the goal is to achieve competition on price to enhance static efficiency, rather than competition on technologies to enhance dynamic efficiency. Put another way, it makes it clear that the goal of the authors of the report is to reject the general rule of intellectual property protection and elevate and expand exceptions and limitations on intellectual property protection to the point that it has no force or effect.

This type of absolutist position is evident where it is stated that "newer drugs will be protected by patent from low-cost competition for at least 20 years, which means that impoverished populations may (and generally will) be deprived of these medicines for that entire period." It ignores the fact that much of a patent term (between 8 and 12 years) expires prior to commercialization in the first market. It fails to acknowledge that medicines currently under development, which will be the first ones globally introduced, already have patent terms that are expiring, and counterparts of these patents may not even exist in the Least Developed Countries. And it glosses over much of what was discussed during the proceedings in the working group and the commissioned paper by Wilder and Solovy (2004) concerning the role of intellectual property and the way in which properly functioning intellectual property systems can be made to work so as to support the emergence of new drugs, and their being taken up for generic manufacture at the end of the patent term.

Finally, it assumes a static, one-dimensional view of patents that does not admit the proper role of patents and the possibilities of the implementation of a strong patent system with built-in safeguards to prevent abuse of patents once granted.

Misperceptions about drug quality assurance and safety

In addition to improving access to medicines, the working group must accept that cooperation is necessary to ensure that medicines are both safe and effective—and that no double standard between developed and developing countries should be tolerated. To do otherwise would be to promote practices and standards below those expected in the developed world, which would clearly be at odds with the basic human rights espoused as one of the principal recommendations of the report.

Safety in medicines has two basic aspects: minimizing the potential for harm while maximizing the potential for benefit, such as efficacy. In this context, two elements are fundamental: first, we should start with the goal of raising the standard of medicines quality for everyone, not lowering the standard for some. Second, while we know some necessary steps have been taken to improve the availability of "approved" drugs, it must be understand that, while necessary, these steps are not sufficient to ensure their quality.

Here are the facts: we know that medicine safety and efficacy are a direct result of implementing quality manufacturing practices. These practices, in turn, have implications for manufacturing costs and ultimately price; those connections and consequences are undeniable and unavoidable. Furthermore, as originator firms scale up to production-level manufacturing, some economies of scale will drive the per-unit cost down. However, many medicines are simply difficult to manufacture from a scientific or technical standpoint. Those manufacturing challenges, with their associated costs, are not diminished simply by an increased scale of production. Unfortunately, this reality results in the pursuit of manufacturing practices at odds with both safety and effectiveness—manifest in both substandard and counterfeit manufacturing operations—with disastrous results for patients.

As indicated in the report, evidence is building that the extent of problems associated with medicines safety and efficacy (such as second- and third-world manufacturing quality) is large and growing in the developing world. Whether due to poor local or regional manufacturing practices or the manufacture or importation of counterfeit products, the results are the same. Furthermore, if we wait for absolute confirmation of this alarming trend, the proof will be the thousands of individuals who have needlessly suffered or died for lack of successful treatment or because of the insidious, silent threat of microbial resistance. The working group is resoundingly silent on this issue.

The critical situation is further complicated by what we see as a considerable misunderstanding in the report about the role of the WHO in drug regulation and approval. The WHO does not "approve" drugs. That is the role of national legal and regulatory authorities that are equipped to do such work. In its prequalification system, the WHO lists those drugs it has prequalified for procurement by UN agencies. As the WHO system stands, there are no requirements to ensure that the drugs it prequalifies meet the standards of safety and efficacy established by national drug regulatory agencies in developed countries. The WHO is clear that it does not guarantee the drugs it prequalifies by issuing a disclaimer on all of its prequalified antiretroviral products, stating that they are "not warranted for safety and/or efficacy in the treatment of HIV/AIDS."

In point of fact, a closer examination of detailed information from the WHO website for the prequalification system and that of the U.S. Food and Drug Administration and the European Medical Evaluation Agency highlight some significant differences vis-a-vis: 1) required clinical trials; 2) basis for legal and regulatory authority; 3) postmarketing surveillance; 4) use of scientific standards for innovator, generic, and copy drugs; and 5) basis for enforcement authority.

The research-based pharmaceutical and biotech industry affirm their strong support for rigorous product marketing reviews and manufacturing protocols that meet the highest international standards for product quality, safety, and efficacy as embodied in the International Conference of Harmonization (ICH). In our view, the WHO prequalification scheme, based on its own record, does not meet the stringent and well established standards in use among leading, highly experienced, and well resourced regulatory agencies. We note that in recent months there has been a confusing series of listings and delistings that have disrupted supply and created confusion for consumers.

With respect to the safety, efficacy, and quality of the medicines purchased by the U.S. government for treating AIDS patients, the President's Emergency Plan for AIDS Relief explains that medicines will be "procured from reliable manufacturers," and supports capacity to test products in the countries in which medicines are delivered. In the specific case of fixed-dose combinations (that is, the combination in a single pill of previously approved individual HIV/AIDS therapies), the U.S. Department of Health and Human Services recently announced an expedited approval process by the Food and Drug Administration that tests efficacy and safety and makes such medicines (including those of foreign companies) available for procurement by the U.S. government. As explained by the U.S. Global AIDS Coordinator and the Secretary of Health and Human Services, "drug patent issues that apply in developed nations should not impede purchase of these drugs for developing countries."[2]

It should also be noted that combination copy products are among those withdrawn from the WHO prequalification list.

At this writing, we observe that the number of products approved by the prequalification scheme that have been subsequently delisted and withdrawn from the market exceeds the products that remain on the prequalified list. Hence it is puzzling to us why the working group strikes a strong note of support for the prequalification scheme approach as a tool to increase access—access yes, but at what cost?

Conclusion

We appreciate the opportunity to provide the working group with a statement of our concerns focused on these four critical thematic areas of the report.

Our industry remains committed to dialogue with all members of the working group and we welcome additional opportunities to explain our perspective as work moves to transforming recommendations into a specific action agenda. We believe that our expertise and experience, while not reflected in the report itself, remains a critically important element in ensuring the successful promotion of medicines access under real life conditions in the field. We stand ready to contribute to the ultimate aim that we all share: health for all and making a real difference in the lives of the world's poor.

Gail Cassell
Vice President, Scientific Affairs
Eli Lilly and Company

Susan Crowley
Director of International Organization Relations
Merck and Co., Inc.

Bill Looney
Senior Director, Global Policy
Corporate Policy–Corporate Affairs
Pfizer, Inc.

Appendix 3

Summary of recommendations

Area	National recommendations	International recommendations
General principles	• Translate principles of human rights relating to medicines access into enforceable rights for the individual. • Make it a priority to strengthen healthcare systems, including the role of ministries of health, capacity building, and integrating public and private sector perspectives. • Explicitly recognize in policies the need to have gender-responsive policies and plans that can be measured effectively (such as using both sex- and gender-disaggregated data) for the extent to which gender is adequately considered and the outcomes substantively target women and girls and increase their equitable access to healthcare and medicines. • Protect and promote women's equal rights, with priority areas being protection from violence, equality under inheritance laws, and increased policy and financial responses to gender-based discrimination in accessing healthcare and treatment.	• International organizations need to acknowledge access to medicines as a human right. • International organizations should support a competitive international pharmaceutical environment that includes generic competition. • UN agencies and the Global Fund to Fight AIDS, Tuberculosis, and Malaria (GFATM) should adopt policies and approaches that ensure that gender considerations are adequately integrated into all aspects of their planning, activities, and budgets.
Barriers to availability		
Gaps in innovation	• Provide policy, sustained funding, and infrastructure for biomedical research, including research on indigenous medicines, to encourage innovation driven by priority health needs. • Promote a policy environment that protects complementary and synergistic roles of publicly and privately funded research. • Promote a predictable, expeditious regulatory environment that emphasizes interventions for priority diseases. • Promote research and development for indigenous medicines.	• Promote global public investment in research for priority health needs of developing countries. • Create an environment that stimulates the private sector to contribute to innovation in public health priorities. • Ensure that international standards for ethical research, such as those elaborated by the Declaration of Helsinki, are applied in all countries. • Request the WHO Commission on Intellectual Property Rights, Innovation, and Public Health to examine alternative international models to the current patent-based system for priority-setting and financing of R&D.

Area	National recommendations	International recommendations
Barriers to availability (continued)		
Unreliable supply systems	• Promote all effective supply channels (public, private, NGO), giving priority to sustainable, reliable supply systems. • Provide clear regulations for supply systems using international best practices, such as those established by the GFATM. • Ensure that the judicial system enforces regulations and supports concrete actions against corruption and diversion. • Explore pooled procurement options.	• Promote transparent information sharing on successful national and pooled supply strategies to enable access. • Provide producers with reliable forecasts of priority product requirements. • Promote international standards for procurement agencies. • Provide technical assistance (WHO, GFATM, the World Bank, and others) to strengthen supply systems in developing countries.
Unsafe medicines	• Strengthen medicines regulatory authority with political support, financing, and staff. • Strengthen national systems for monitoring adverse reactions. • Work with appropriate parties to monitor adverse drug events and institute no-fault systems for redressing drug injury. • Design national registration systems that are simple, work in a timely manner, and include fast tracks for priority medicines; in particular, those that are WHO prequalified.	• Enforce compliance with international GMP. • Share information about poor-quality products and producers, based on reliable and accurate data. • Strengthen systems for sharing information on benefit-risk assessment and regulatory decisions (such as withdrawals of medicines from the market). • Support existing adverse event monitoring. • Prequalify according to the WHO prequalification project and monitor priority products and suppliers and share this information (for example, as a "white list"). • Strengthen and expand the WHO prequalification project and make it a permanent and well funded function of the WHO. • Strengthen national regulatory capacity through training, capacity building, information sharing, evaluation of best practices, and sustained funding.
Barriers to affordability		
Inadequate and unfair financing	• Increase the public sector budget for essential medicines and ensure equitable access. • Phase out user fees for essential medicines in favor of more equitable financing. • Institute performance-based payment for providers. • Promote good donation practices as in accord with international guidelines.	• Increase total international financing for essential medicines targeting the poor. • Do not use loans to fund recurrent medicines expenditures. • Increase development assistance for health in line with recommendations by the Commission on Macroeconomics and Health and others. This assistance should support national health policies and systems. • Coordinate and simplify donor assistance procedures to make transparent the total funding flows and to reduce transaction costs for developing countries.
Unaffordable prices	• Use a variety of tools to lower prices in developing countries ("equity pricing") including promoting generic competition, using essential medicines lists, promoting bulk procurement, negotiating equitable prices, minimizing markups, and adapting national legislation to ensure that TRIPS safeguards can be used. • Ensure timely payment of suppliers to encourage lower prices.	• Carefully monitor the impact of TRIPS compliance by medicine-producing countries, such as India and China, on access to essential medicines in developing countries and present an assessment by 2007. • Provide medicines at production cost to low-income countries and at reduced cost to middle-income developing countries. • Share accurate and consistent data on producer prices, markups and profits, tariffs and taxes, and fees and other charges. • Support a competitive international pharmaceutical environment that includes generic competition. • Ensure that regional and bilateral trade negotiations promote international understandings that support access to medicines.

Area	National recommendations	International recommendations
Barriers to appropriateness		
Inappropriate prescribing and dispensing	• Implement national coordinating policy on activities to improve rational medicines use. • Use evidence-based treatment guidelines in teaching, monitoring, and evaluation. • Ensure responsible and ethical medicines promotion by pharmaceutical companies through government oversight. • Ensure the availability of independent and impartial information for continuing education of prescribers and dispensers. • Separate prescribing and dispensing profits. • Train, regulate, and monitor people prescribing and dispensing medicines.	• International donor agencies should coordinate to support country efforts to promote rational use. • WHO should update and promote ethical criteria for medicines promotion and medicines information (for example by utilizing the Internet). • WHO should share, disseminate, and translate independent information on treatment of priority conditions for national adaptation.
Inappropriate use by households	• Promote culturally appropriate health literacy and community support. • Ensure availability of independent and impartial information for households using culturally appropriate means. • Mobilize and engage communities to improve use of medicines. • Regulate consumer advertising for medicines.	• International health organizations and NGOs should continue to develop and disseminate health literacy information related to appropriate use of medicines for use in developing countries.
Cross-cutting issues		
Human resources	• Pay health workers an adequate wage. • Ensure sufficient numbers of trained pharmacy workers of different levels. • Continuously update and adapt to needs training curricula for prescribers and dispensers. • Develop, support, and involve the communities' own resource persons.	• Support health worker training with updated international curricula. • Use donor financing to fund salary costs in poorest countries. • Institute international agreements and cooperation on health worker migration. • Through international financing agencies such as the World Bank and the GFATM and major bilateral donors, focus on training and capacity building of a substantial number of supply chain managers and other essential health workers in developing countries.
Gender	• Ensure women have access to accurate, gender-sensitive medicines information. • Involve women in medicines policymaking. • Promote innovative and outcome-based research on the gendered aspects of medicines access and use by women and girls and men and boys. • Collect sex- and gender-disaggregated data on access and use, which, in combination with adequate gender analysis, should inform policies, plans, and budgets. • Ensure that women and girls have equal access to medicines. • Ensure full and equitable access to sexual and reproductive health services and commodities. • Ensure that national essential medicines lists contain the core medicines and devices for sexual and reproductive health recommended by the UNFPA and WHO.	• Ensure that adequate gender analysis is included in all health policymaking, strategies, and programs.
Institutional structures	• Create and maintain efficient national institutions required to implement law, regulation, inspection, and financing in the field of medicines.	• Maintain international institutions capable of supporting the development of stable medicines access systems.

Notes

Executive summary

1. In this report, the terms *medicines* and *pharmaceuticals* will be regarded as equivalent, referring both to therapeutic agents and to vaccines. The term *drugs* is ambiguous since it is often applied to substances causing addiction and prone to misuse, many with little or no significance in medical treatment; it will be generally avoided in the present report except in direct quotation. The Working Group on Access to Essential Medicines has noted that certain other products are closely analogous to medicines and deserve similar approaches, for example, the intrauterine contraceptive device.

2. The working group's framework is in part based on WHO's four-part framework to describe the main elements that affect access to essential medicines:

- Rational selection and use.
- Affordable prices.
- Sustainable financing.
- Reliable health and supply systems.

It is the basis for the development and implementation of WHO's Department of Essential Drugs and Medicines Policy (EDM) work to increase access to medicines. This framework has been adopted by WHO's main partners (WHO 2004a). The working group, in carrying out its analysis, dealt with these elements in detail.

3. Equity pricing is a concept launched by the WHO in the late 1990s. It is based on the ethical notion that developing countries should not be asked to pay for medicine development costs, marketing, and shareholder returns. This is a much wider concept than differential pricing and encompasses all the active policy and administrative measures a government or procurement organization can take to achieve differential pricing related to purchasing power. These measures include price information and transparency, pooled procurement, reduction of taxes and margins, price negotiations, voluntary licensing agreements, and, as an ultimate measure, compulsory licensing. Equity pricing is the political choice and action, differential pricing is the result. Equity pricing has been successfully practiced for more than 30 years for children's vaccines and reproductive health commodities.

Chapter 1

1. In this report, the terms *medicines* and *pharmaceuticals* will be regarded as equivalent, referring both to therapeutic agents and to vaccines. The term *drugs* is ambiguous since it is often applied to substances causing addiction and prone to misuse, many with little or no significance in medical treatment; it will be generally avoided in the present report except in direct quotation. The Working Group on Access to Essential Medicines has noted that certain other products are closely analogous to medicines and deserve similar approaches, for example, the intrauterine contraceptive device.

2. The working group also relied on the Essential Medicines Concept, which is defined as follows:

> Essential medicines are those that satisfy the priority health care needs of the population. They are selected with due regard to public health relevance, evidence on efficacy and safety, and comparative cost effectiveness. Essential medicines are intended to be available within the context of functioning health systems at all times in adequate amounts, in the appropriate dosage forms, with assured quality and adequate information, and at a price the individual and the community can afford. The implementation of the concept of essential medicines is intended to be flexible and adaptable to many different situations; exactly which medicines are regarded as essential remains a national responsibility (WHO 2004a).

3. See also the Global Alliance for TB Drug Development for additional information on MDR-TB, [www.tballiance.org/2_1_2_MDR_TB.asp].

4. Writing in January 2001, Attaran and Sachs concluded that, in order to contain AIDS, aid would need to be increased within the succeeding three years to a minimum of $7.5 billion or more; they pointed out that this sum could easily be afforded by the OECD donor countries.

5. For an authoritative definition of a generic medicine, see Laurence and Carpenter (1998). Essentially the term is applied to a medicine that is not (or is no longer) protected by patent and is being supplied by a manufacturer other than the originator, generally under an international nonproprietary name. Usage differs somewhat: some definitions limit use of the term to those products that have been certified by national regulatory agencies as being fully bioequivalent to the original patented product. Others apply the term to medicines that are shown to be essentially equivalent, but for which testing for bioequivalence has not been done.

6. It is useful to review the definitions and difference between the terms *gender* and *sex*. They are not synonyms, and using them interchangeably obscures the problem and limits the scope of response. The term *gender* is used to describe those characteristics of women and men that are socially constructed; the term *sex* refers to those that are biologically determined. People are born female or male but learn to be girls and boys who grow into women and men. This learned behavior contributes to defining gender identity and will largely determine gender roles in any specific time and cultural setting. This in turn will be evident across the range of human experience including, for example, how medicines are accessed and used differently by men and women.

7. The program to eradicate smallpox, primarily based on vaccination, cost more than $300 million over the whole of its 12-year life but saved hundreds of millions of dollars per year in directly measurable costs.

8. Brazil's ambitious program to counter AIDS has had striking economic results. The investments made have paid off in terms of savings, such as in the costs of hospitalization and in the purchase of patented medicines at world prices. The Ministry of Health has

estimated that the universal free provision of AIDS medicines prevented 234,000 AIDS-related hospital admissions during the period 1997–2000, saving $677 million for the country's health system.

9. It has been estimated that about a third of the rural poverty in China is caused by catastrophic medical spending; the majority of medical expenses in China relate to medicines (Evidence provided by Dr. Yuanli Liu to the working group during 2003 meeting in Geneva).

10. A 1992 study showed that per capita expenditure on medicines at that time ranged from $412 in Japan to $2 or less in Bangladesh and parts of Sub-Saharan Africa. At the middle of the range, per capita expenditure was $97 in the United Kingdom and $89 in Norway (Ballance and others 1992). At the end of the century these discrepancies had still not been reduced (Bannenberg 2000; Scrip 2000; WHO 2000b, 2000d).

11. A long-term research program maintained in India by the Ciba-Geigy company was abandoned some 20 years ago. A more recent effort is covered by an agreement between Merck Inc. and the Costa Rican Biodiversity Institute (InBIO) in 1994.

12. The current definition reads: "Essential medicines are those that satisfy the priority health care needs of the population. They are selected with due regard to public health relevance, evidence on efficacy and safety, and comparative cost effectiveness. Essential medicines are intended to be available within the context of functioning health systems at all times in adequate amounts, in the appropriate dosage forms, with assured quality and adequate information, and at a price the individual and the community can afford. The implementation of the concept of essential medicines is intended to be flexible and adaptable to many different situations; exactly which medicines are regarded as essential remains a national responsibility" (WHO 2004a).

13. More information is available on the prequalification website (http://mednet3.who.int/prequal/).

14. The "3 by 5" goal represents a commitment by WHO. It is not a separate program or fund, but an objective that it hopes to attain through a series of mutually complementary measures (WHO 2003a). It must be borne in mind that the total number of individuals infected is currently estimated at 42 million, though not all require intensive medicinal treatment.

15. In a suitably competitive market, generic medicines are as a rule much less expensive than their originator counterparts. Researchers from the Hudson Institute have countered that generics AIDS medicines are more expensive (Adelman, Norris, and Weicher 2004). It would appear that the authors misinterpreted MSF data; at the least, more research is needed. Clearly prices are very dynamic, so it entirely possible that a certain group of medicines, in a certain context, will be priced differently than the general rule.

16. See also www.haiweb.org/medicinesprices for pricing data being generated by country studies using the WHO and HAI pricing survey methodology. These findings also tend to support the general view that generics are less expensive.

17. Manufacturing in these countries is often severely hampered by the fact that virtually all starting materials and equipment have to be imported and that their turnover is small; as a result they can have difficulty in competing in terms of cost and quality with generics manufacturers abroad. The future significance of national manufacturing in low-income countries will need to be carefully considered. Arguments for maintaining at least some of these facilities include the following advantages:

- The ability to produce simple bulk products (such as intravenous fluids) where international transport costs can be prohibitive.
- The ability to produce traditional medicines.
- The value of a production plant as an educational and research center, such as to facilitate pharmacy training.

Chapter 2

1. The working group's framework is in part based on WHO's four-part framework to describe the main elements that affect access to essential medicines:

1. Rational selection and use.
2. Affordable prices.
3. Sustainable financing.
4. Reliable health and supply systems.

It is the basis for the development and implementation of WHO's Department of Essential Drugs and Medicines Policy work to increase access to medicines. This framework has been adopted by WHO's main partners. The working group, in carrying out its analysis, dealt with these elements in detail.

2. The mainstays of malaria treatment have been chloroquine and sulfadoxine-pyrimethamine, both of which are available at negligible cost. The former is no longer effective against *Plasmodiun falciparim* in most tropical areas, however, and resistance to the latter is now widespread. There are certain alternative drugs, but they are currently too expensive for entire populations (see White 1999).

3. Sleeping sickness, after markedly decreased incidence in the 1960s, is becoming increasingly more prevalent in many Sub-Saharan African countries. The disease causes at least 40,000 deaths annually. Armed conflicts in many of the endemic areas, as well as the focused international attention on other major infectious diseases has contributed to this resurgence. An overview compiled by MSF in 1999 found that the supply of all four applicable medicines had either ceased (eflorinthine Hcl) or become insecure (suramin Na, melarsoprol, and pentamidine isethionate) because of lack of commercial interest in production (Pécoul and others 1999).

4. The fact that these wealthy nations are themselves increasingly challenging the prices of medicines and seeking lower cost solutions for their domestic markets must be noted, but falls outside the scope of the present report.

5. This is currently the situation in Afghanistan, where the Avicenna Institute was re-equipped for large-scale medicine production in order to ensure national self-sufficiency. Production has largely ceased and the future of the institute as a manufacturing center is under review (Graham Dukes, personal communication, 2003). A former Western pharmaceutical factory (built by the Hoechst company of Germany) is still mothballed, but the Business Humanitarian Forum, a partnership founded in collaboration with research-based companies, and partnered with the European Generics Association, hopes to reestablish production in the country.

6. In Viet Nam alone, it is estimated that 3 million people fall below the poverty line each year because of health-related expenditure (see Wagstaff and van Doorslaer 2003).

7. It is important to emphasize that the discussions of intellectual property rights for medicines in this report are not an attack on the entire patent system. The challenge is to ensure that patents are not a barrier in increasing access to affordable essential medicines for the poor in developing countries. Prices afforded by patent protection should not be a barrier to affordability and availability in these countries. To this end, certain TRIPS flexibilities exist, and other alternatives— such as how to promote voluntary licensing and technology transfer—need to be discussed constructively.

8. An example of the governmental use clause in practice is provided by Cameroon, a developing country member of the African Intellectual Property Organization (OAPI). The organization grants regional patents that are valid in all OAPI member states, and a significant number of antiretrovirals are currently protected by OAPI patents. Some of the patented agents are, however, available at lower prices from generic sources. In order to make the best possible use of its limited resources, the Ministry of Health of Cameroon

authorized the public procurement agency in 2000 to buy antiretrovirals from generic sources.

9. The governments of Kenya and Brazil, in announcing programs to supply generic medicines for the treatment of AIDS, have referred to the epidemic as a national emergency in their countries.

10. There is a clear market failure for medicines in developing countries. The hope that [intellectual property] protection would provide a financial incentive to drug firms to invest in drugs for tropical diseases has not materialized; during the last decade, research and development for developing country diseases has declined rather than increased. In 1975–99 only 1 percent of 1,191 new medicines approved for marketing were specifically indicated for a tropical disease. Poor countries do not constitute a market capable of inducing patent-driven investment (Lehman 2002). The global market for pharmaceuticals was estimated at $406 billion in 2002, with the US, EU, and Japan accounting for 80 percent of this market and the rest of the world combined for only 20 percent (IMS Health 2001).

11. As part of a three-pronged approach to AIDS, the Brazilian government decided in 1996 to make the necessary medicines available free of charge to those who needed them. (Brazil, Ministry of Health 2001: Law 9,313 of 13 November 1996). Fourteen antiretroviral drugs are currently available in this way. This has been possible because of the other aspects of the national medicines policy involving price negotiations with the suppliers, the threat of compulsory licensing, and training in medicines use. It is notable that Brazil has used the threat of compulsory license successfully in price negotiations but has never actually needed to issue a compulsory license. The Brazilian AIDS program is heavily subsidized by the government in terms of finance and staffing, and this model would be difficult to transfer to poorer countries with higher AIDS burdens without additional and long-term donor support.

12. The two leading guides on pricing are the MSH–WHO International Drug Price Indicator Guide and the UNICEF–MSF–WHO list of sources and prices of selected medicines and diagnostics for HIV/AIDS.

13. Reductions of 35 percent or more are often cited, but much greater reductions are sometimes achieved. In mid-2003, the pool representing the countries of the Andean region agreed with producers of antiretroviral medicines on drastically reduced prices; the price for one three-component product, which had been as high as $5,000 per person per year in one participating country, fell to $365 for all 10 countries.

14. In Argentina and Brazil, for example, the agencies for a long period accepted so-called *similares*—that is, secondary versions of medicines for which equivalence with the original product had been demonstrated only in vitro (such as in disintegration testing of tablets).

15. A broadly constituted meeting convened by MSF and the Drugs for Neglected Diseases Initiative in Geneva in July 2003 concluded, "There are substantial fears that some ICH guidelines might have a negative impact on access to essential medicines in developing countries. Specifically, new stringent requirements for raw materials may raise drug prices without offering any discernible public health benefit in exchange. Some medicines that are badly needed in developing countries may not be granted regulatory approval, since risk/benefit calculations are necessarily made differently in non-ICH and ICH countries. In addition, the existing governance structure excludes many of the stakeholders affected by the process, including developing countries, consumers, and health professionals. . . . the motivation behind extending the guidelines beyond ICH countries is not clear. Higher standards for the quality of raw materials and drugs may allow ICH countries to protect themselves from lower-priced (generic) imports from other markets that do not hold to

ICH quality standards, while at the same time ensuring continued access to high quality raw materials from non-ICH countries for their domestic manufacturers" (MSF 2001).

16. There are a small number of medicines for which the toxic dose is only slightly higher than the dose normally used in treatment (such as digitalis, which is used for heart disorders). For these substances and their pharmaceutical forms, exceptionally high standards have to be maintained.

17. For example, many developing countries import generic products from India. At their best, Indian medicines have been found to be of a standard at least equivalent to that of the equivalent originator medicines. A persistent problem is the complex and inadequate system of official inspection of manufacturing plants, partly as a consequence of the division of responsibility between the federal and state authorities. As a result, some manufacturing sources remain well below acceptable standards of quality assurance (Dukes 2001). Many developing countries therefore find it necessary to apply strict batch quality control to products of Indian origin, or prefer to purchase such medicines through a nonprofit intermediary capable of exercising its own quality control procedures.

18. The problem of substandard antimalarials has been particularly well documented in western Kenya.

19. During an epidemic of meningitis involving 41,000 cases in Niger in 1995, the country was promised a donation of 88,000 vaccine doses from Nigeria, with Pasteur Mérieux and SmithKline Beecham as manufacturers. In fact the vaccines were found to have been replaced on the way with spurious copies containing no active ingredient but with labeling meticulously copied from the original (Pinel and others 1997).

20. See examples from the Philippines, Pakistan, and Nigeria cited by Velásquez, Madrid, and Quick (1998).

21. The handbook of guidelines is available on multiple sites, including www.med.rug.nl/pharma/who-cc/ggp/homepage.htm.

22. http://dcc2.bumc.bu.edu/prdu/HTML_DOCS_TOC.htm.

23. http://dcc2.bumc.bu.edu/prdu/Session_Guides/effective_public_education.htm.

Chapter 3

1. A notable example cited to the working group is the long-term success of the national Essential Drugs Programme in Bhutan, dating from 1987. Whereas medicines access was earlier very limited, it is estimated that 90 percent of the population now enjoys access to high-quality essential medicines. In 1995, retail prices were on average 6 percent lower than in 1985, and the prices paid by the program in the course of procurement are currently some 50 percent below world market prices. Monitoring is intensive, with facilities reporting twice yearly on their stocks and use of medicines; only 0.75 percent of the overall budget was wasted as a result of medicines expiry (Stapleton 2000).

2. The case of Chad illustrates both successes and failures. Under a national medicines policy adopted in 1998, the proportion of the population with access to essential medicines rose from 46 percent in 1999 to 60 percent in 2001, and annual public expenditure on medicines was trebled over a six-year period, though it still amounted to only $0.12 per capita. The average percentage of essential medicines available in health facilities fell from 80 to 70 percent over the same period, while the average duration of stock-outs increased from 41 to 59 days. Standard treatment guidelines were updated, but no improvements were recorded in the use of antibiotics or in the (excessive) use of injections (WHO 2001).

3. The rapid flow of donor funds for health and medicines into Uganda after the fall of the Amin regime and the restoration of a democratic regime with which donors could

work constructively provides a classic example. The results can now be assessed after 15 years of experience (DMFA 1995).

4. In Timor-Leste shortly after independence, the number of physicians available to serve a population of 800,000 was negligible. Practical nurses with elementary training were handling both diagnosis and treatment. A specially adapted handbook was therefore devised to assist them in performing this task as well as possible without the need for frequent referral (Dukes, field report 2001).

In many developing countries, poorly paid health workers earn money by levying charges on the medicines they issue to patients. Rather than attempting to prohibit this practice, it sometimes proves better to regularize it so that modest fixed charges (insufficient in themselves to constitute a barrier to access) can be made in order to provide the health worker with a living wage.

5. Good examples of successful handbooks of this type include those from Ghana, Uganda, and Zimbabwe.

6. In one African state in 2003, a delegation from a Western aid foundation proposed that it provide a massive sum of money from Western countries to be used for the purchase of AIDS medicines. Consultation led to the conclusion that the population would be better served if the proposed supply were somewhat reduced and a fair proportion of the total sum were used instead to provide training in the management, diagnosis, and treatment of the disorder. Medicines would otherwise not be used to the best advantage (Dukes, personal communication 2003).

Chapter 4

1. Equity pricing is a concept launched by WHO in the late 1990s. It is based on the ethical notion that developing countries should not be asked to pay for medicine development cost, marketing, and shareholder returns. This is a much wider concept than differential pricing and encompasses all the active policy and administrative measures a government or procurement organization can take to achieve differential pricing related to purchasing power. These measures include price information and transparency, pooled procurement, reduction of taxes and margins, price negotiations, voluntary licensing agreements, and, as an ultimate measure, compulsory licensing. Equity pricing is the political choice and action; differential pricing is the result. It has been successfully practiced for more than 30 years for children's vaccines and reproductive health commodities.

Appendix 1

1. Diseases associated with widespread changes in eating and exercise patterns, such as those associated with migration from rural to urban settings.

2. www.health.go.ug and www.cdc.gov.

3. www.health.go.ug/health_units.htm.

4. Every year, in order to estimate the level of access to essential medicines, the WHO Action Programme on Essential Drugs interviews relevant experts in each country about the pharmaceutical situation. The interviewees can choose from four levels of access to essential medicines by the population: less than 50 percent, 50–80 percent, 80–95 percent, and above 95 percent. They indicate which category is most appropriate for their country. Essential medicines are those that satisfy the healthcare needs of the majority of the population.

5. www.kaisernetwork.org/daily_reports/rep_index.cfm?hint=1&DR_ID=24203.

6. www.uppap.or.ug/.

Appendix 2

1. World Health Organization. 2004. "Access to HIV/AIDS Drugs and Diagnostics of Acceptable Quality: Procurement Quality and Sourcing Project." Geneva. Available online at http://mednet3.who.int/prequal/.

2. "HHS Proposes Rapid Process for Review of Fixed Dose Combination and Co-Packaged Products." May 16, 2004. [www.hhs.gov/news/press/2004pres/20040516.html].

References

Abbott, F. M. 2002. "Compulsory Licensing for Public Health Needs: The Trips Agenda at the WTO after the Doha Declaration on Public Health." Occasional Paper 9, Friends World Committee for Consultation. Quaker United Nations Office, Geneva. [www.geneva.quno.info/pdf/OP9%20Abbott.pdf].

Adelman, C. C. 2002. "The Full Measure of Foreign Aid." In U.S. Agency for International Development, *Foreign Aid in the National Interest: Promoting Freedom, Security, and Opportunity.* Washington, D.C.: USAID.

———. 2003. "The Privatization of Foreign Aid: Reassessing National Largesse." *Foreign Affairs* 82(6): 9–14.

Adelman, C. C., J. Norris, and S. J. Weicher. 2004. "Myths and Realities on Prices of AIDS Drugs." White Paper. Hudson Institute, Washington, D.C. [www.hudson.org/files/publications/AdelmanARVWhitePaper.pdf].

Angell, M. 2004. *The Truth About the Drug Companies.* New York: Random House.

Annan, K. 2001. "Road Map Towards the Implementation of the United Nations Millennium Declaration." Follow-up to the outcome of the Millennium Summit. Report of the Secretary-General. United Nations Document A/56/326. New York. [www.un.org/documents/ga/docs/56/a56326.pdf].

Attaran, A., and J. Sachs. 2001. "Defining and Redefining International Donor Support for Combating the AIDS Pandemic." *The Lancet* 357: 57–61.

Autier, P., R. Govindaraj, R. Gray, R. Laksh, R. Minarayanan, H. G. Nassry, and G. Schmets. 2002. *Drug Donations in Post-emergency Situations.* World Bank HNP Discussion Paper. Washington, D.C.

Baker, B. 2003. "Analysis and Response to WTO Action Regarding Paragraph 6 of the Doha Declaration on the TRIPS Agreement and Public Health." Paper commissioned by the Working Group on Access to Medicines, UN Millennium Project Task Force on HIV/AIDS, Malaria, TB, and Access to Essential Medicines. UN Millennium Project, New York.

Ballance, R., J. Pogany, H. Forstner. 1992. *World's Pharmaceutical Industries: An International Perspective on Innovation, Competition, and Policy.* London: UNIDO (United Nations Industrial Development Organization) and Edward Elgar, Inc.

Bannenberg, W. 2000. "Action on Essential Drugs." Presentation at the 13th International AIDS Conference, July 10–14, Durban, South Africa.

———. 2002. "ARV Parallel Export from Africa to Europe." Posting on E-drug-digest, October 7. [www.essentialdrugs.org/edrug/archive/200210/msg00023.php].

———. 2004. "Impact of ICH on Access to Essential Medicines in Developing Countries." Presentation at the Neglected Diseases Group meeting, February 6–7, Penang, Malaysia.

Banta, H. D. 2001. "Worldwide Interest in Global Access to Drugs." *Journal of the American Medical Association* 285(22): 2844–46.

Bill & Melinda Gates Foundation. 2002. "Developing Successful Global Health Alliances." Seattle, Wash. [www.gatesfoundation.org/nr/downloads/globalhealth/GlobalHealthAlliances.pdf].

Bisilliat, J. 2001. "Introducing the Gender Perspective in National Essential Drug Programmes." Department of Essential Drugs and Medicines Policy. World Health Organization, Geneva.

Black, R. E. 2003. "Where and Why are 10 Million Children Dying Every Year?" *The Lancet* 361: 2220–34.

Boatin, B. 2003. "Rapid Expansion of Access to ARVs: The Litmus Test of Drug Management for the 21st Century." Paper presented at the Strategies for Enhancing Access to Medicines Conference December 10–12, New York. [www.msh.org/seam/conference/presentations/Agenda/AgendaDay1/transcripts/3%20-%20Boakye%20Boatin.pdf].

Brazil, Ministry of Health. 2001. Ministry of Health of Brazil, National AIDS Drug Policy. Brasilia. [www.aids.gov.br/assistencia/aids_drugs_policy.htm]

Broun, D., and G. Dukes. 1993. "National Drug Policies and Use of Resources." World Bank Briefing Paper. Washington, D.C.

Buse, K., and G. Walt. 2000. "Global Public-Private Partnerships for Health, Parts I and II." *Bulletin of the World Health Organization* 78: 549–61; 699–709.

Byström M., and P. Einarsson. 2001. "TRIPS: Consequences for Developing Countries: Implications for Swedish Development Cooperation." Consultancy Report to the Swedish International Development Cooperation Agency (SIDA). Stockholm. [www.grain.org/docs/sida-trips-2001-en.PDF].

Caines, K., J. Bataraingaya, L. Lush, G. Murindwa, and H. N'jie. 2003. "Impact of Public-Private Partnerships Addressing Access to Pharmaceuticals in Low Income Countries: Uganda Pilot Study." Initiative on Public-Private Partnerships for Health, Geneva.

Coleman, P. 2003. "The Cost-effectiveness of Artemisinin Combination Therapies." Presentation at The Institute of Medicine London's Economics of Antimalarial Drugs workshop, March 13, London.

CMH (Commission on Macroeconomics and Health). 2001. *Final Report of the Commission on Macroeconomics and Health.* World Health Organization. Geneva. [www.cmhealth.org/].

CIPR (United Kingdom Commission on Intellectual Property Rights). 2002. *Integrating Intellectual Property Rights and Development Policy.* London. [www.iprcommission.org/graphic/documents/final_report.htm].

Correa, C. 2002. "Implication of the Doha Declaration on the TRIPS Agreement and Public Health." Health Economics and Drugs Department, Essential Drugs and Medicines Series Paper 12. World Health Organization, Geneva. [www.who.int/medicines/library/par/who-edm-par-2002-3/doha-implications.doc].

Delacollette, C., P. Van der Stuyft, and K. Molima. 1996. "Using Community Health Workers for Malaria Control: Experience in Zaire." *Bulletin of the World Health Organization* 74(4): 423–30.

De Vries, Th. P. G. M., R. H. Henning, H. V. Hogerzeil, D. A. Fresle. 1995. "Guide to Good Prescribing." World Health Organization, Geneva. [www.med.rug.nl/pharma/who-cc/ggp.htm].

DFID (U.K. Department for International Development). 2004a. *Access to Medicines in Underserved Markets: What Are the Implications of Changes in Intellectual Property Rights, Trade and Drug Registration Policy?* United Kingdom Government Policy and Plans. London. [www.dfid.gov.uk/pubs/files/dfidsynthesispaper.pdf].

———. 2004b. *Increasing Access to Essential Medicines in the Developing World.* United Kingdom Government Policy and Plans. London. [www.dfid.gov.uk/Pubs/files/accessmedicines.pdf].

Deininger, K., P. Mpuga. 2004. "Economic and Welfare Effects of the Abolition of Health User Fees: Evidence from Uganda." World Bank and Policy Research Working Paper Series, number 3276.

Denmark, Ministry of Foreign Affairs. 1995. *Danish Ministry of Foreign Affairs: Review and Appraisal Report, Essential Drugs Management Programme, Uganda.* Euro Health Group A/S and Uganda Essential Drugs Programme. Copenhagen.

Drahos, P., and J. Braithwaite. 2002. "Intellectual Property, Corporate Strategy, Globalisation: TRIPs in Context." *Wisconsin International Law Journal* 20: 451–80.

Dukes, M.N.G., ed. 1993. *Drug Utilization Studies: Methods and Uses.* WHO Regional Office for Europe, WHO Regional Publications: European Series 45. Geneva: World Health Organization.

———. 2001. "Report to the World Bank on the Drugs Inspectorate in Uttar Pradesh." Mimeo. World Bank, Washington, D.C.

———. 2003. "Theoretical Model of a National Drug Policy." Diagram, Masters Programme in International Community Health. University of Oslo, Oslo.

Dye, C., B. Williams, M. A. Espinal, and M. C. Raviglione. 2002. "Erasing the World's Slow Stain: Strategies to Beat Multidrug-Resistant Tuberculosis." *Science* 295(5562): 2042–46. [www.tballiance.org/2_1_2_MDR_TB.asp].

Edwards, N. 1996 "Molecule of the Month: Taxol." University of Bristol, Molecule of the Month series. Bristol, U.K. [http://www.bristol.ac.uk/Depts/Chemistry/MOTM/taxol/taxol.htm].

Ekman, B. 2004. "Community-based Health Insurance in Low-income Countries: A Systematic Review of the Evidence." *Health Policy and Planning* 19(5): 249–70.

European Consultation on the Rights of Patients and WHO (World Health Organization). 1994. "Declaration on the Promotion of Patients' Rights in Europe." Proceedings of the European Consultation on the Rights of Patients, March 28–30, Amsterdam. [www.who.int/genomics/public/eu_declaration1994.pdf].

Everard, M. 2003. "The AIDS Medicines and Diagnostics Facility." Presentation to the SEAM/Rockefeller Symposium, November 10–11, New York.

Folb, P. 2004. "New Drugs for Developing Countries: Obstacles and Opportunities." Paper commissioned by the Working Group on Access to Medicines, UN Millennium Project Task Force on HIV/AIDS, Malaria, TB, and Access to Essential Medicines. UN Millennium Project, New York.

Fraser, H. S. F., D. Jazayeri , P. Nevil , Y. Karacaoglu, P. E. Farmer, E. Lyon, M. K. Smith-Fawzi, F. Leandre, S. Choi , and J. S. Mukherjee. 2004. "An Information system and medical record to support HIV treatment in rural Haiti." *British Medical Journal* 329(7475): 1142–46.

Gillespie-White, L. 2001. "What Did Doha Accomplish? Assessing the Doha Declaration on Intellectual Property Rights and Access to Medicines." International Intellectual Property Institute, Washington, D.C. [http://mail.iipi.org/db/views/detail.asp?].

GFHR (Global Forum for Health Research). 2004. "Helping Correct the 10/90 Gap." Website. [http://www.globalforumhealth.org/pages/index.asp].

Guilloux, A., and S. Moon. 2000. "Hidden Price Tags: Disease-Specific Drug Donations: Costs and Alternatives." Médecins Sans Frontières, Geneva. [www.accessmed-msf.org/prod/publications.asp?scntid=492001217138&contenttype=PARA&].

Haddad, W. F. 2004. "Generic Medicines, the Solution or the Problem?" Paper prepared for the Working Group on Access to Essential Medicines, UN Millennium Project Task Force on HIV/AIDS, Malaria, TB, and Acecss to Essential Medicines. UN Millennium Project, New York.

Hardon, A. 2001. " Immunisation for All? A Critical Look at the First GAVI Partners Meeting." *HAI Europe HAI Lights* 6(1):2–9.

Hafeez, A., and Z. Mirza. 1999. "Responses from Pharmaceutical Companies to Doctors' Requests for More Drug Information in Pakistan: Postal Survey." *British Medical Journal* 319(7209): 547.

Health Action International, Médecins Sans Frontières, and Consumer Project on Technology. 1999. "Amsterdam Statement to WTO Member States on Access to Medicines." Paper prepared at the Increasing Access to Essential Drugs in a Globalised Economy: Working Towards Solutions conference, November 25–26, Amsterdam. [www.accessmed-msf.org/upload/ReportsandPublications/392001512503/Amsterdam%20statemt.pdf].

Hill, S., and K. Johnson. 2004. *Emerging Challenges and Opportunities in Drug Registration and Regulation in Developing Countries.* DFID HSRC (Department for International Development Health Systems Resource Center). London. [http://www.dfidhealthrc.org/Shared/publications/Issues_papers/ATM/Hill.pdf].

't Hoen, E. 2000. "TRIPS, Pharmaceutical Patents, and Access to Essential Medicines: A Long Way from Seattle to Doha." *Chicago Journal of International Law* 3(1): 27–46.

Hogerzeil, H. V. 2003. "Access to Essential Medicines as a Human Right." *Essential Drug Monitor* 33: 25–26. [www.who.int/medicines/library/monitor/33/EDM33_25-26_Access_e.pdf].

Hogerzeil, H. V., Bimo, D. Ross-Degnan, R. O. Laing, D. Ofori-Adjei, B. Santoso, A. K. Azad Chowdhury, A. M. Das, K. K. Kafle, A. F. B. Mabadeje, and A. Y. Massele. 1993. "Field Tests for Rational Drug Use in Twelve Developing Countries." *The Lancet* 342(8884): 1408–10.

Hogerzeil, H. V., M. R. Couper, and R. Gray. 1997. "Guidelines for drug donations." *British Medical Journal* 314(7082): 737.

Huttin, C. 1994. "The Chinese Medicines Market: Moving towards a Market System?" *Health Policy* 29(3): 247–59.

IAPCG (Interagency Pharmaceutical Coordination Group). 2000. "Report on Activities in 1996–1999." WHO/UNICEF/UNFPA/World Bank. [http://www.who.int/medicines/organization/par/ipc/introipc.shtml].

IFPMA (International Federation of Pharmaceutical Manufacturers Associations). 2003. Website. www.ifpma.org.

———. 2004. *Research and Development for Neglected Diseases: Lessons Learned and Remaining Challenges.* Geneva. [www.ifpma.org/documents/NR1897/RDNeglected-Diseases.pdf].

IMS Health. 2001. *Market Report.* [www.ims-global.com].

———. 2002. "IMS World Review" Website. [www.ims-global.com/products/sales/review.htm].

INRUD (International Network for Rational Use of Drug). 2002. Newsletters 11(1) and 11(2). [http://www.inrud.org/news_pdf/vol11no1_nopics.pdf and http://www.inrud.org/news_pdf/vol11no2_nopics.pdf].

Irwin, A., and E. Ombaka. 2003. "Background Paper of the Task Force on Major Diseases and Access to Medicines, Subgroup on Access to Essential Medicines." UN Millennium Project, New York. [http://www.unmillenniumproject.org/documents/tf05atemapr18.pdf].

Irvine, B. 2004. "Death and Taxes: The Impact of Taxes and Other Barriers to Access." Campaign for Fighting Disease. [http://www.fightingdiseases.org/pdf/taxes-tariffs-access.pdf].

Jarrett, S. 2003. "UNICEF Bulk Procurement of Vaccines and Pharmaceuticals." Presentation to the SEAM/Rockefeller Symposium, November 10–11, New York.

Kaplan, W., R. Laing, B. Waning, and L. Levison. 2003a. "Pharmaceutical Regulation and Quality Assurance." Unpublished review and discussion paper prepared for RPM-Plus. Boston University, Boston.

———. 2003b. "The Impact of Regulatory Interventions on Pharmaceutical Access and Quality: What Is the Evidence and Where Are the Gaps in our Knowledge?" Unpublished paper by the Boston University School of Public Health, Boston. [http://dcc2.bumc.bu.edu/richardl/RPM+_Project/Regulation&QA/Regulation-evidence-Feb2003.doc].

Kaul, I. 2000. "The Fairness Revolution: Global Public Goods: A New Way to Balance the World's Books." *Le Monde Diplomatique,* June.

Kawasaki, E., and J. Patten. 2002. "Drug Supply Systems of Missionary Organizations: Identifying Factors Affecting Expansion and Efficiency: Case Studies from Uganda and Kenya." WHO #HQ/01191467. World Health Organization, Geneva. [http://dcc2.bumc.bu.edu/richardl/ih820/Resource_materials/MEDS_JMS/Default.htm].

Kim J., J. Millen, A. Irwin, and J. Gershman, eds. 2000. *Dying for Growth: Global Inequality and the Health of the Poor.* Monroe, Maine: Common Courage Press.

Kibumba, G. 2002 "Human resource capacity in pharmaceutical sector." On-line discussion group posting Thursday, August 22, 2002. E-Drug. [http://www.essentialdrugs.org/edrug/archive/200208/msg00015.php].

Kinney, E. R. 2001. "The International Human Right to Health." *Indiana Law Review* 34: 1457–75.

Laing, R. 2001. "Health and Pharmacy Systems in Developing Countries." Paper presented at the WHO/WTO Workshop on Differential Pricing and Financing of Essential Drugs, April 8–11, Høsbjør, Norway. [www.wto.org/english/tratop_e/trips_e/hosbjor_presentations_e/05laing_e.pdf].

Laing, R., H. V. Hogerzeil, and D. Ross-Degnan. 2001. "Ten Recommendations to Improve Use of Medicines in Developing Countries." *Health Policy and Planning* 16(1): 13–20.

Laing R., A. Waning, A. Gray, N. Ford, and E. 't Hoen. 2003. "25 years of the WHO Essential Medicines Lists: Progress and Challenges." *The Lancet* 361: 1723–29.

Laurence, D. R., and J. R. Carpenter. 1998. *A Dictionary of Pharmacology and Allied Topics.* Amsterdam: Elsevier Science Ltd.

Leon, D. A., and G. Walt, eds. 2000. *Poverty, Inequality, and Health: An International Perspective.* New York: Oxford University Press.

Lehman, B. 2001. "Patents and Health." Discussion paper prepared for the WIPO Policy Advisory Commission Task Force meeting, May 22, Beijing.

———. 2003. "The Pharmaceutical Industry and the Patent System." International Intellectual Property Institute. Washington, D.C. [www.earthinstitute.columbia.edu/cgsd/documents/lehman.pdf].

Leufkens H., F. Haaijer-Ruskamp, A. Bakkar, and G. Dukes. 1994. "Scenario Analysis of the Future of Medicines." *British Medical Journal* 309 (6962): 1137–40.

Lewis-Lettington, R., and C. Banda. 2004. *A Survey of Policy and Practice on the Use of Access to Medicines-related TRIPS Flexibilities in Malawi.* Access to Medicines issues paper. DFID Health Systems Resource Center. London. [www.dfidhealthrc.org/Shared/publications/Issues_papers/ATM/Lettington.pdf].

Light, D.W., and J. Lexchin. 2004. "Will Lower Drug Prices Jeopardize Drug Research?" *American Journal of Bioethics* 4(1): W1–W4.

Lilly. 2003. Website. "Building Healthier Societies Through Partnership (Updated May 2004)." [www.lillyMDR-TB.com].

Liu, Y. 2003. "Towards Systematic Analysis and Policy Recommendations for ATEM." Presentation to the Working Group on Access to Essential Medicines, United Nations Millennium Project Task Force on HIV/AIDS, Malaria, TB, and Access to Essential Medicines, New York.

Love, J. 2000. "Frequently Asked Questions about Compulsory Licensing." *International Journal of Risk and Safety in Medicine* 13: 2.

Lunde, P. K. M. 1984. "WHO's Programme on Essential Drugs: Background, Implementation, Present State and Perspectives." *Dan Med J* 311ff.

Mamdani, M. 1992. "Early Initiatives in Essential Drugs Policy." In N. Kanji, A. Hardon, J. M. Harnmeijer, M. Mamdani, and G. Walt, eds., *Drugs Policy in Developing Countries.* London: Zed Books.

Maponga, C., and C. Ondari. 2003. "The Quality of Antimalarials—A Study in Selected African Countries." EDM Research Paper WHO/EDM/PAR/2003.4. World Health Organization, Geneva.

McGuaran, A. 2002. "Royal College Issues New Guidelines On Gifts From Drugs Companies." *British Medical Journal* 325(7363): 511.

Mnyika, K. S., and J. Z. Kilewo. 1991. "Irrational Drug Use in Tanzania." *Health Policy Planning* 6(2): 180–184.

MSF (Médecins Sans Frontières). 2001. *Fatal Imbalance—The Crisis in Research and Development for Drugs for Neglected Diseases.* Report by the MSF Access to Essential Medicines Campaign and the Drugs for Neglected Diseases Working Group. Geneva. [www.msf.org/source/access/2001/fatal/fatalshort.pdf].

———. 2002a. Circular letter from the MSF-sponsored Campaign for Access to Essential Medicines, July 24, Geneva.

———. 2002b. "From Durban to Barcelona—Overcoming the Treatment Deficit." Geneva. [www.doctorswithoutborders.org/publications/reports/2002/fdtb_07-2002.pdf].

———. 2003a. *Doha Derailed: A Progress Report on TRIPS and Access to Essential Medicines.* Briefing for the 5th World Ministerial Conference, September 10–14, Cancun. [http://www.accessmed-msf.org/prod/publications.asp?scntid=27820031110385&contenttype=PARA&].

———. 2003b. "New Tricks to Limit Access to Affordable Medicines." *MSF Access* 9: 2 [www.msf.org.au/docs/newsletters/access_9.pdf].

———. 2003c. "Surmounting Challenges: Procurement of Antiretroviral Medicines in Low- and Middle-Income Countries in Low and Middle Income Countries: the Experience of MSF." Geneva. [www.accessmed-msf.org/documents/procurementreport.pdf].

———. 2003d. "Trading Away Health." *MSF Access* 9: 2. [www.msf.org.au/docs/newsletters/access_9.pdf].

———. 2003e. "Untangling the Web of Price Reductions: A Pricing Guide for the Purchase of ARVS for Developing Countries." [http://www.accessmed-msf.org/documents/untangling4thapril2003.pdf].

———. 2004a. *Running Out of Breath: TB Care in the 21st Century.* Geneva [www.msf.org/source/annual/tbday/2004/TBDay2004report.doc].

———. 2004b. *Untangling the Web of Price Reductions: A Pricing Guide for the Purchase of ARV's for Developing Countries.* Geneva.

MSH (Management Sciences for Health) and WHO. 2000. *Managing Drug Supply (Second Edition).* West Hartford, Conn.: Kumarian Press. [http://erc.msh.org/mainpage.cfm?file=1.0.htm&module=drugs&language=English].

———. 2002. "International Drug Price Indicator Guide." Management Sciences for Health, and World Health Organization, Boston and Geneva. [http://erc.msh.org/mainpage.cfm?file=1.0.htm&id=1&temptitle=Introduction&module=DMP&language=English].

Moss, N. E. (2002). "Gender equity and socioeconomic inequality: A framework for the patterning of women's health." *Social Science and Medicine* 54: 649–61.

Mossialos, E., and G. Dukes. 2001. "Affordably Priced New Drug for Poor Populations: Approaches for a Global Solution." *International Journal of Risk and Safety in Medicine* 16(1): 1–29.

Mubyazi, G. M., and G. Hutton. 2003. "Understanding Mechanisms for Integrating Community Priorities in Health Planning, Resource Allocation and Service Delivery: Results of a Literature Review." Equinet Discussion Paper 13. Harare, Zimbabwe. [www.equinetafrica.org/bibl/docs/mechanisms.pdf].

Mukherjee, J., P. Farmer, F. Leandre, W. Lambert, M. Raymonville, S. Koenig, D. Walton, M. Colas, P. Nevil, N. Louissant, and C. Orilus, 2003. "Assessing the Success of an Integrated HIV Prevention and Care Project in Rural Haiti." World Health Organization Case Study. Geneva. [www.ahfgi.org/global_pdf/haiti_study.pdf].

Newton, P. N., N. J. White, J. A. Rosendaal, and M. D. Green. 2002. "Murder by Fake Drugs." *British Medical Journal* 324(7341): 800–801.

NIHCM (U.S. National Institute for Health Care Management). 2000. "Prescription Drugs and Intellectual Property Protection." NIHCM Foundation Issue Brief. Washington, D.C.

OHCHR (Office of the UN High Commissioner for Human Rights) and UNAIDS. 2002. "HIV/AIDS and Human Rights: International Guidelines. Revised Guideline 6: Access to Treatment, Care, Prevention, and Support." United Nations, New York. [www.unaids.org/html/pub/Publications/IRC-pub02/JC905-Guideline6_en_pdf.pdf].

Ombaka, E. and A. Irwin. 2002. Background Paper for the Access to Essential Medicines Working Group, Task Force on HIV/AIDS, Malaria, TB, and Access to Essential Medicines, UN Millennium Project.

Osmond, B. 2002. "A Brief Description of the Revolving Drug Fund of the Khartoum Comprehensive Child Care Programme." Save the Children Fund Discussion Paper. London.

Oswald, W., E. Leontsini, and G. Burnham. 2004. "Innovative Practices for Improved Access to Essential Medicines: Women, Poverty, and the Role of Community in the Control of Disease." Paper commissioned by the Working Group on Access to Essential Medicines, UN Millennium Project Task Force on HIV/AIDS, Malaria, TB, and Access to Essential Medicines. UN Millennium Project, New York.

Oxfam. 2002. "Generic competition, price and access to medicines: The case of antiretrovirals in Uganda." Oxfam Briefing Paper 26. [http://www.oxfam.qc.ca/html/politique/PDFpolitique/26generic.pdf].

———. 2003. "From Cancun to Miami": The FTAA Threat to Development in the Hemisphere." [www.oxfamamerica.org/newsandpublications/publications/briefing_papers/art6446.html].

———. 2004. "TRIPS-Plus Provisions." [www.oxfamamerica.org/whatwedo/issues_we_work_on/trade/news_publications/trips/art5391.html].

Pang, T., M. A. Lansang, and A. Haines. 2002. "Brain Drain and Health Professionals: A Global Problem Needs Global Solutions." *British Medical Journal* 324: 499–500.

PAHO (Pan American Health Organization). 2004. "Report on the 38th Session of the Subcommittee on Planning and Programming." 8th Session of the Subcommittee on Planning and Programming, March 24–26, Washington, D.C. [www.paho.org/English/GOV/CE/ce134-05-e.pdf].

Pavin, M., T. Nurgozhin, G. Hafner, F. Yusufy, and R. Laing. 2003. "Prescribing Practices of Rural Primary Health Care Physicians in Uzbekistan." *Tropical Medicine and International Health* 8(2): 182–190.

Pécoul, B., P. Chirac, P. Troullier, and J. Pinel. 1999. "Access to Essential Drugs in Poor Countries—A Lost Battle?" *Journal of the American Medical Association* 281(4): 361–67.

Pinel, J., F. Varaine, F. Fermon, G. Marchant, and G. Marioux. 1997. "Des faux vaccins anti-meningocoque lors d'une epidemie de meningite au Niger." *Med Maladies Infect.* 1997; 27:1–563.

Preker, A. S., G. Carrin, D. Dror, M. Jakab, W. Hsiao, and D. Arhin-Tenkorang. 2002. "Effectiveness of Community Health Financing in Meeting the Cost of Illness." *Bulletin of the World Health Organization* 80:143–50. [www.scielosp.org/pdf/bwho/v80n2/a10v80n2.pdf].

Quality Assurance and Safety of Medicines. 2003. "Prequalification Project." Website. [http://mednet3.who.int/prequal/about.shtml]

Quick, J. D. 2003. "Ensuring Access to Essential Medicines in the Developing Countries: A Framework for Action." *Clinical Pharmacology and Therapeutics* 73(4): 1–6.

Raufu, A. 2002. "Nigerian Health Authorities Worry Over Exodus of Doctors and Nurses." *British Medical Journal* 325 (7355): 65.

RCP (Royal College of Physicians) 2004. "The Relationship Between Physicians and the Biomedical Industries: Advice of the Royal College of Physicians." Advice of the Royal College. London. [www.rcplondon.ac.uk/college/statements/advice_biomedIndustry.htm].

Rankin, R. J. 2003. "Procurement Options and Issues." Presentation at the SEAM/Rockefeller Symposium, November 10–11, New York.

Republic of Uganda. 2002. "Uganda Second Participatory Poverty Assessment Process"(final report). Kampala. [www.uppap.or.ug/nationalPPA2.htm].

Richter, J. 2004. *Public-Private Partnerships and International Health Policy Making. How Can Public Interests be Safeguarded?* Development Policy Information Unit, Elements for Discussion Series. Ministry of Foreign Affairs of Finland, Helsinki.

Sachs, J., J. McArthur, G. Schmidt-Traub, C. Bahadur, M. Faye, and M. Kruk. 2004. "Millennium Development Goals Needs Assessments for Bangladesh, Cambodia, Ghana, Tanzania, and Uganda". UN Millennium Project background paper. New York. [www.unmillenniumproject.org/documents/mp_ccspaper_jan1704.pdf].

Scherer, F. M., and J. Watal. 2001. "Post-TRIPS Options for Access to Patented Medicines in Developing Countries." Commission on Macroeconomics and Health Working Paper Series. Paper for Working Group on Access to Essential Medicines, UN Millennium Project Task Force on HIV/AIDS, Malaria, TB, and Access to Essential Medicines. Cambridge, Mass.

Schultz, T. P. 1989. "Returns to Women's Education." PHRWD Background Paper 98/001. World Bank, Washington, D.C.

Scrip. 1977. "Scrip World Pharmaceutical News." 9(259) PJB Publications, Richmond, Va.

———. 2000. *Scrip's Yearbook, 2000.* Richmond, Va.: PJB Publications.

SEAM (Strategies for Enhancing Access to Medicines). 2003. Presentations at the SEAM/Rockefeller Foundation Symposium on Pooled Procurement, November 10–11, New York.

Snell, B. 1999. "Inappropriate Drug Donations: The Need for Reform." *The Lancet* 358(9281): 578–80.

Stapleton, M. 2000. "Bhutan Essential Drugs Programme; A Case History." WHO/EDM/DAP/2000.2. World Health Organization, Geneva.

Stilwell, B., and. N. Awofeso. 2004. "Managing Brain Drain and Brain Waste of Health Workers in Nigeria." *Bulletin of the World Health Organization* 82: 595–600. [www.who.int/bulletin/bulletin_board/82/stilwell1/en/].

Stone, L. 1986. "Primary Health Care for Whom: Village Perspectives from Nepal." *Social Science and Medicine* 22(3): 293–302.

Sunartono, D. 1995. "From Research to Action: the Gunungkidul Experience." *Essential Drugs Monitor* 20: 21–22.

Taylor, Y., ed. 2004. *Battling HIV/AIDS. A Decision Makers Guide to the Procurement of Medicines and Related Supplies.* Washington D.C.: World Bank [http://siteresources.worldbank.org/INTPROCUREMENT/Resources/Technical-Guide-Procure-HIV-AIDS-Meds.pdf].

Trap, B., and C. Lessing. 1995. *ZEDAP (Zimbabwe Essential Drugs Action Programme) Survey 1995.* Zimbabwe Ministry of Health and Welfare, Harare.

Trap, B., E. Holme Hansen, and H. V. Hogerzeil. 2002. "Prescription Habits of Dispensing and Non-dispensing Doctors in Zimbabwe." *Health Policy and Planning* 17(3): 268–96.

Troullier P., P. Olliaro, E. Torreele, J. Orbinski, R. Laing, and N. Ford. 2002. "Drug Development for Neglected Diseases: A Deficient Market and a Public Policy Failure." *The Lancet* 359: 2188–94.

Uganda. 2001. "2001 Demographic and Health Survey." Uganda Bureau of Statistics. Entebbe.

———. 2002a. "Second Uganda Participatory Poverty Assessment Report: Deepening the Understanding of Poverty." Ministry of Finance, Planning and Economic Development. Entebbe.

———. 2002b. "Uganda Pharmaceutical Sector Baseline Survey. Financial and Technical Assistance from HAI and WHO. Ministry of Health. Entebbe.

UK Commission. 2002. "Report of the United Kingdom Commission on Intellectual Property Rights." London.

UNAIDS. 2002. *Report on the Global HIV/AIDS Epidemic 2002.* Geneva.

UNCESCR (United Nations Committee on Economic, Social, and Cultural Rights). 2001. "Substantive Issues Arising in the Implementation of the International Covenant on Economic, Social and Cultural Rights." Statement adopted at the Third United Nations Conference on the Least Developed Countries, May 14–20, Brussels. [www.unctad.org/en/docs/aconf191bp_7.en.pdf].

UNFPA (United Nations Population Fund) and WHO (World Health Organization). 2003. "Essential Drugs and Other Commodities for Reproductive Services." Draft Discussion Document. United Nations, New York. [www.unfpa.org/upload/lib_pub_file/198_filename_commodities_rhs.pdf].

UNHCHR (United Nations Office of the High Commission on Human Rights). 1966. "International Covenant on Economic, Social, and Cultural Rights, Article 12(1)." Geneva. [http://www.unhchr.ch/html/menu3/b/a_cescr.htm].

UNIDO–AAITPC. 2001. "Pharmaceutical and Health Products' Sector Profile." United Nations Industrial Development Organization Asia-Africa Investment and

Technology Promotion Center. [www.unido-aaitpc.org/unido-aaitpc/new1/Uganda/pharmaceuticals.pdf].

UN Millennium Project. 2005a. *Trade for Development.* Report of the Task Force on Trade. Sterling, Va.: Earthscan.

———. 2005b. *Investing in Strategies to Reverse the Global Incidence of TB.* Report of the Task Force on HIV/AIDS, Malaria, TB, and Access to Essential Medicines, Working Group on TB. Sterling, Va.: Earthscan.

———. 2005c. *Investing in Development: A Practical Plan to Achieve the Millennium Development Goals.* Sterling, Va.: Earthscan.

United Nations. 1948. *Universal Declaration of Human Rights,* Art. 25(1). UN Resolution 217 A (III). New York.

———. 2000. "United Nations Millennium Declaration." General Assembly Resolution 55/2. New York. [www.un.org/millennium/declaration/ares552e.pdf].

USP (United States Pharmacopeia) and DQI (Drug Quality Information Program). 2004. "A Review of Drug Quality in Asia with Focus on Anti-Infectives." Washington, D.C. and Rockville, Md. [http://www.uspdqi.org/pubs/other/ANEReview.pdf].

Van der Geest, S. 1992. "Village Health Workers as Medicine Sellers?" *International Journal of Health Planning and Management* 7 (2): 97–99.

Velásquez, G., Y. Madrid, and J. D. Quick. 1998. "Health Reform and Drug Financing: Selected Topics." Economics and Financing document WHO/DAP/98.3. WHO, Geneva.

Vivas Eugui, D. 2003. "Regional and Bilateral Agreements and a TRIPS-plus World: The Free Trade Area of the Americas." The Quaker United Nations Office, International Centre for Trade and Sustainable Development, Quaker International Affairs Programme, Geneva and Ottawa. [www.geneva.quno.info/pdf/FTAA%20(A4).pdf].

Vlassoff, C. and C. G. Moreno. 2002. "Placing gender at the centre of health programming: challenges and limitations." *Social Science and Medicine* 54: 1713–23.

Wagstaff, A., and E. van Doorslaer. 2003. "Catastrophe and Impoverishment in Paying for Health Care: With Applications to Vietnam 1993–98." *Health Economics* 12(11): 921–34.

Walt, G. 1990. *Community Health Workers in National Programs: Just Another Pair of Hands?* Philadelphia: Open University Press.

WHA (World Health Assembly). 2001. "Resolution WHA 55.14 of the 54th World Health Assembly on accessibility of essential medicines." WHO, Geneva.

———. 2003. "Resolution 56.27 of the 56th World Health Assembly on Intellectual Property Rights, Innovation and Public Health." 56th World Health Assembly on Intellectual Property Rights, Innovation, and Public Health, May 19–28, Geneva.

White, N. J. 1999. "Antimalarial Drug Resistance and Mortality in *Falciparum* Malaria." *Tropical Medicine and International Health* 4(7): 469–470.

WHO (World Health Organization). 1977. "Essential Medicines." Technical Report Series 615. Geneva.

———. 1985. *The Rational Use of Drugs.* Report of the Conference of Experts, November 25–29, Nairobi.

———. 1988. "Ethical Criteria for Medicinal Drug Promotion." Geneva. [www.who.int/medicines/library/dap/ethical-criteria/ethicalen.shtml].

———. 1990. *The Role of the Pharmacist in the Health Care System.* Document WHO/PHARM/DAP/90.1. Geneva.

———. 1997. *Public-Private Roles in the Pharmaceutical Sector.* Document WHO/DAP/97.12. Geneva.

———. 1998a. *Estimating Drug Requirements—A Practical Manual.* Geneva.

———. 1998b. "Health Reform and Drug Financing." Drug Action Programme Document WHO/DAP/98.3. Geneva.

———. 1999a. "Guidelines for Drug Donations." 2nd edition. WHO/EDM/PAR/99.4. Geneva. [www.who.int/medicines/library/par/who-edm-par-1999-4/who-edm-par-99-4.shtml].

———. 1999b. "The World Medicines Situation." Geneva.

———. 2000a . "Country Progress Indicators for Components of WHO Medicines Strategy 2000–2003." In WHO, *WHO Medicines Strategy: Framework for Action in Essential Drugs and Medicines Policy.* Geneva.

———. 2000b. "The Essential Drugs Strategy." Department of Essential Drugs and Medicines Policy. Geneva.

———. 2000c. "Framework for Action." *WHO Policy Perspectives on Medicines* 2(1): 2.

———. 2000d. *Global Comparative Pharmaceutical Expenditures with Related Reference Information.* Health Economics and Drugs, EDM Series 3. Document EDM/PAR/2000.2. Geneva.

———. 2000e. "Major Progress, Yet Huge Inequities Remain." *WHO Policy Perspectives on Medicine* 2(1): 1

———. 2000f. "WHO Medicines Strategy 2000–2003: Framework for Action in Essential Drugs and Medicines Policy 2000–2003." Geneva. [www.who.int/medicines/strategy/strategy2000_2003.shtml].

———. 2001. "Essential Drugs and Medicines Policies: Achievements and Constraints: 2000–2001." PowerPoint presentation. Geneva.

———. 2002a. *25 Questions and Answers on Health and Human Rights.* Health and Human Rights Publication Series 1. Geneva.

———. 2002b. "Access to Essential Medicines Is Fundamental to Human Rights." *WHO Policy Perspectives on Medicines* 4: 5.

———. 2002c. *Core Indicators for Monitoring National Drug Policies.* Geneva.

———. 2002d. "Global Strategy for Traditional and Complementary Alternative Medicine 2002–2005." Geneva. [Retrieved November 22, 2004, from www.who.int/medicines/library/trm/trm_strat_eng.pdf].

———. 2002e. "How to Develop and Implement a National Drug Policy, (Second Edition)." *WHO Policy Perspectives on Medicines* 6. Geneva. [www.who.int/medicines/library/edm_general/6pagers/No6-6pg-en.pdf].

———. 2002f. "Promoting Rational Use of Medicines: Core Components." *WHO Policy Perspectives on Medicines* 5:1. Geneva.

———. 2002g. *World Health Report 2002: Reducing Risks, Promoting Healthy Life.* Geneva. [www.who.int/whr/2002/en/].

———. 2003a. "3 by 5 Programme for HIV/AIDS." WHO Fact Sheet 274. Geneva.

———. 2003b. *Framework for Action in Essential Drugs and Medicines Policy 2004–2007.* Document WHO/EDM/2004.2. Geneva. [www.who.int/medicines/strategy/strategy2004_2007.shtml].

———. 2003c. "How to Develop and Implement a National Drug Policy." *WHO Policy Perspectives on Medicines* 6:1. Geneva. [www.who.int/medicines/library/edm_general/6pagers/No6-6pg-en.pdf].

———. 2003d. "Substandard and Counterfeit Medicines." WHO Factsheet #275. Geneva. [www.who.int/mediacentre/factsheets/fs275/en/].

———. 2003e. *WHO Model Formulary.* Geneva.

———. 2003f. *The World Health Report, 2003: Shaping the Future.* Geneva. [www.who.int/whr/2003/en/].

———. 2004a. "The World Medicines Situation." Geneva. [Retrieved December 1, 2004, from www.who.int/medicines/organization/par/World_Medicines_Situation.pdf].

———. 2004b. "Priority Medicines for Europe and the World: A Public Health Approach to Innovation" Section 6.8: "Tuberculosis." Geneva. [http://mednet3.who.int/prioritymeds/report/background/tuberculosis.doc].

———. 2004c. "Uganda." Geneva. [Retrieved December 30, 2004, from www.who.int/hac/about/donorinfo/en/Uganda%20v7-SC-CD-KS.pdf].

———. 2004d. "International Nonproprietary Names." Geneva. [www.who.int/medicines/organization/qsm/activities/qualityassurance/inn/orginn.shtml].

———. 2004e. *The World Health Report 2004: Changing History.* Geneva. [www.who.int/whr/en/].

WHO (World Health Organization) and HAI (Health Action International). 2003. "Medicine Prices—A New Approach to Measurement." World Health Organization and Health Action International. Geneva. [www.who.int/medicines/library/prices.shtml].

WHO (World Health Organization) and IFPMA (International Federation of Pharmaceutical Manufacturers and Associations). 2000. "Improving Access to Essential Drugs through Innovative Partnerships: Antimalarials for Sub-Saharan Africa." Action Paper. Geneva.

———. 2001. "Working Paper on Priority Infectious Diseases Requiring Additional R&D." Paper prepared for the Round Table Meeting of WHO/Industry Drug Development Working Group, July 2001. [www.who.int/intellectualproperty/documents/en/IFPMA.pdf].

WHO (World Health Organization) and MSH (Management Sciences for Health). 2004. "Drugs and Therapeutics Committee: A Practical Guide." WHO/EDM/PAR/2004.1. Geneva. [www.who.int/medicines/library/par/who-edm-par-2004_1/WHO_EDM_PAR_2004_1_Drugs_and_therapeutics_committees.pdf].

WHO (World Health Organization) and TDR. (2003). *The Behavioral and Social Aspects of Malaria and its Control* (TDR/STR/SEB/VOL/03.1). Geneva: World Health Organization.

WHO (World Health Organization) and WTO (World Trade Organization). 2001. "Final Report." Workshop on Differential Pricing and Financing of Essential Drugs, April 8–11, Hosbjor, Norway. [www.wto.org/english/tratop_e/trips_e/hosbjor_report_e.pdf].

Wilder, R. and E. M. Solovy. 2004. "The Role of Intellectual Property in Procurement and Development of Medicines for Developing Country Diseases." Background paper prepared for the Working Group on Access to Medicines, UN Millennium Project Task Force on HIV/AIDS, Malaria, TB, and Access to Essential Medicines. UN Millennium Project, New York.

World Bank. 1993. *World Development Report 1993: Investing in Health.* New York: Oxford University Press. [www-wds.worldbank.org/servlet/WDSContentServer/WDSP/IB/1993/06/01/000009265_3970716142319/Rendered/PDF/multi0page.pdf].

———. 2004a. "HIV/AIDS Medicines and Related Supplies: Contemporary Context and Procurement Technical Guide." Washington, D.C.: World Bank.

———. 2004b. World Development Indicators Online (by subscription).

WTO (World Trade Organization). 1994. *TRIPS: Agreement on Trade-Related Aspects of Intellectual Property Rights. PART II—Standards Concerning the Availability, Scope and Use of Intellectual Property Rights.* Geneva. [www.wto.org/english/tratop_e/trips_e/t_agm3_e.htm].

———. 2001. "Declaration on the TRIPS Agreement and Public Health." Resolution adopted at the Doha WTO Ministerial Meeting, November 14, Doha. [www.wto.org/english/thewto_e/minist_e/min01_e/mindecl_trips_e.htm].

———. 2003. "Implementation of Paragraph 6 of the Doha Declaration on the TRIPS Agreement and Public Health." Decision of the General Assembly, Document WT/L/540. Geneva. [www.wto.org/english/tratop_e/trips_e/implem_para6_e.htm].

WTO Council for Trade-Related Aspects of Intellectual Property Rights. 2001. "Special Discussion on Intellectual Property and Access to Medicines 4." WTO Doc No IP/C/M/31. Geneva.

Yamey, G. 2002. "Public Sector Must Develop Drugs for Neglected Diseases." *British Medical Journal* 324: 698.

Yemen. 1996. *Statistical Report of the Supreme Board of Drugs and Medical Appliances.* Sanaa.

Zakus, J., and C. Lysack. 1998. "Review Article: Revisiting Community Participation." *Health Policy and Planning* 13: 1–12.

Zammit, A. 2003. *Development at Risk: Re-Thinking UN-Business Partnerships.* The South Center and the United Nations Research Institute for Social Development. Geneva. [www.southcentre.org/publications/developmentatrisk/newrisk.pdf].

Task Force on Hunger
Halving hunger: it can be done

Task Force on Education and Gender Equality
Toward universal primary education: investments, incentives, and institutions

Task Force on Education and Gender Equality
Taking action: achieving gender equality and empowering women

Task Force on Child Health and Maternal Health
Who's got the power? Transforming health systems for women and children

Task Force on HIV/AIDS, Malaria, TB, and Access to Essential Medicines
Working Group on HIV/AIDS
Combating AIDS in the developing world

Task Force on HIV/AIDS, Malaria, TB, and Access to Essential Medicines
Working Group on Malaria
Coming to grips with malaria in the new millennium

Task Force on HIV/AIDS, Malaria, TB, and Access to Essential Medicines
Working Group on TB
Investing in strategies to reverse the global incidence of TB

Task Force on HIV/AIDS, Malaria, TB, and Access to Essential Medicines
Working Group on Access to Essential Medicines
Prescription for healthy development: increasing access to medicines

Task Force on Environmental Sustainability
Environment and human well-being: a practical strategy

Task Force on Water and Sanitation
Health, dignity, and development: what will it take?

Task Force on Improving the Lives of Slum Dwellers
A home in the city

Task Force on Trade
Trade for development

Task Force on Science, Technology, and Innovation
Innovation: applying knowledge in development